D0857311

# Adsorption Technology
## *for Air and Water Pollution Control*

Kenneth E. Noll
Vassilios Gounaris
Wain-Sun Hou

LEWIS PUBLISHERS, INC.

**Library of Congress Cataloging in Publication Data**

Noll, Kenneth E.
  Adsorption technology for air and water pollution control /
Kenneth E. Noll, Vassilios Gounaris, Wain-sun Hou.
      p.    cm.
  Includes bibliographical references and index.
    1. Air—Pollution. 2. Water—Purification—Adsorption. 3. Gasses—
Absorption and adsorption. I. Gounaris, Vassilios. II. Hou, Wain-sun.
III. Title.
TD883.N55  1991     91-21280
628.5—dc20
ISBN 0-87371-340-0

LEWIS PUBLISHERS, INC.
121 South Main Street, Chelsea, Michigan 48118

Printed in the United States of America 1 2 3 4 5 6 7 8 9 0

# Adsorption Technology
## for Air and Water Pollution Control

# AUTHORS

Kenneth E. Noll is Professor and Chairman of the Pritzker Department of Environmental Engineering at the Illinois Institute of Technology, Chicago, Illinois. He is also Director of the Industrial Waste Elimination Research Center and formerly was Director of the Midwestern Regional Air Pollution Training Center, both of which are sponsored by the USEPA.

Professor Noll is a recognized authority in the field of Air Pollution Control Technology. He is the author of several books on air pollution control and on waste minimization. He received his Ph.D. from the University of Washington and has had previous experience as an Air Force Meteorologist, Senior Engineer with the California Air Resources Board, and Professor of Environmental Engineering at the University of Tennessee.

Dr. Hou is a Consulting Engineer with Patterson Associates, Inc. in Chicago, Illinois where he is involved in projects concerning Industrial Pollution Control. He received his Ph.D. from the Department of Environmental Engineering at the Illinois Institute of Technology.

Mr. Gounaris is currently a Ph.D. candidate at the Illinois Institute of Technology. Both Dr. Hou and Mr. Gounaris have particular research interests in adsorption processes, which are currently widely used for volatile organic control in industry.

# ACKNOWLEDGMENTS

The research presented in this book was funded for ten years (1980-1989) by the United States Environmental Protection Agency - Industrial Waste Elimination Research Center (IWERC) located in the Department of Environmental Engineering at the Illinois Institute of Technology (IIT), Chicago, Illinois. The objective of the research was to develop and integrate Air and Water Adsorption Technology for applications to Pollution Control, Waste Elimination, and Resource Recovery.

Valuable guidance was provided by the IWERC's first Director, Dr. James W. Patterson, and the center's Scientific and Industrial Advisory Committees. Dr. C.N. Haas (IIT) first suggested the application of the differential reactor technique and provided valuable technical assistance. Dr. J.C. Crittenden (Michigan Tech.) and two of his graduate students provided important assistance by supplying computer adsorption models and assisted in their application during the early stages of the research. Dr. M. Satoh constructed the first gravimetric system for Air Adsorption studies. Dr. A.A. Aguwa constructed the first differential reactor system for Liquid Adsorption studies. Dr. M.C. Yeh constructed the first gravimetric system for Liquid Adsorption studies. Dr. E. Furuya provided valuable theoretical support over the entire period of this project both as a visiting scholar and during numerous campus visits.

The research required to develop the information presented in this book was performed by numerous graduate students and research associates from many countries, and involved a team effort. The people who have contributed to this book and their country of origin are as follows:

Visiting Scholars: A. Arai (Japan); E. Furuya (Japan); M. Satoh (Japan). Visiting Researchers: Y.S. Su (China); D.H. Wang (China). Research Assistants: A.A. Aguwa (Nigeria); A. Belalia (Algeria); P.S. Bartholomew (U.S.A.); P.T. Boulanger (France); G. Dinopoulou (Greece); Y.P. Fang (Taiwan); G.J. Ferret (France); V. Gounaris (Greece); W.S. Hou (Taiwan); J. Hsieh (Taiwan), A. Khalili (Iran); B.G. Pierce (U.S.A.); J.N. Sarlis (Greece); T. Shen (China); J.M. Wu (Taiwan); M.C. Yeh (Taiwan).

Because some researchers made major contributions to the development of information for specific chapters, they are given special recognition as Contributing Authors at the beginning of the chapters.

# TABLE OF CONTENTS

# LIST OF FIGURES

# LIST OF TABLES

# CHAPTER I

## ADSORPTION AS A TREATMENT PROCESS

*INTRODUCTION*

Adsorption is a process by which material accumulates at the interface between two phases. These phases can be any of the following combinations: liquid-liquid, liquid-solid, gas-liquid, and gas-solid. The adsorbing phase is called the adsorbent, and any substance being adsorbed is termed as an adsorbate.

Adsorption onto solid adsorbents has great environmental significance, since it can effectively remove pollutants from both aqueous and gaseous streams. Due to the high degree of purification that can be achieved, this process is often used at the end of a treatment sequence.

Activated carbon, the most popular adsorbent, has been traditionally used for the removal of odors, taste, and colors which are caused by trace pollutants. Its high adsorptive capacity and versatility have expanded its application to the treatment of numerous industrial waste streams.

In many of those industrial applications, recovery and reuse of the removed pollutants have become possible, since, under certain conditions, the process is reversible. This feature can convert this rather expensive process into a source of net profit.

A number of synthetic adsorbents having increased reversibility have recently been developed. Although their versatility and adsorptive capacity are generally less than those of carbon, they are advantageous for certain applications. Some commercial adsorbents and their properties are presented in Table 1.1.

The application of adsorption technology for pollution control usually deals with the control of organic compounds. VOC's, pesticides, PCB's, phenolics, and complex synthetic organics are typical adsorbates. In general, any organic compound having molecular weight greater than 45 is likely to be a good adsorbate on activated carbon. It has been observed that most of the non-biodegradable organics are good adsorbates; this makes adsorption very compatible with acti-

Table 1.1.   Physical Properties of Adsorbents

| Material | Shape | Size (Mesh) | $\epsilon_p$ (-) | $\rho_b$ (lb/ft$^3$) | $r_p$ (Å) | $A_s$ (m$^2$/g) | Commercial Products |
|---|---|---|---|---|---|---|---|
| *-Aluminas-* | | | | | | | |
| Active aluminas | G | 0.25-0.3 | 50 | | 35-45 | 235 | Alcoa F-1 Reynolds RA-1, RA-3 |
| | S | 3-8 | 0.5-0.6 | 47-50 | 40-50 | 400 | Alcoa H-151 Kaiser KA-201 |
| | T | 1/8-1 in | 0.3 0.47 | 50 50 | 136 99 | 90 190 | Harshaw A1-0104T A1-1404T |
| CoCl$_2$-impregnated | G | 8-14 | 0.3 | 54 | 45 | 200 | Alcoa F-6 |
| Desiccant | G | 3-8 | 0.3 | 57 | 45 | 200 | Alcoa F-5 |
| Catalytic alumina | S | V | 0.62 | 47 | 45 | 300 | Pechiney CR |
| Activated bauxite | G | 8-20 | 0.35 | 53 | 50 | - | Florite |
| Chromatographic alumina | G | 80-200 | 0.3 | 58 | 45 | 225 | Alcoa F-20 |
| *-Silicas-* | | | | | | | |
| Aluminosilicates | C S P | 4-12 | 0.4 0.55 | 40 55 | - | 770 | Davison 3A, 4A, 5A, 13X, 700 |

*(continued)*

Table 1.1. Physical Properties of Adsorbents *(continued)*

| Material | Shape | Size (Mesh) | $\epsilon_p$ (-) | $\rho_b$ (lb/ft$^3$) | $r_p$ (Å) | $A_s$ (m$^2$/g) | Commercial Products |
|---|---|---|---|---|---|---|---|
| Alumino-silicates | | | | | | | Linde 3A, 4A, 5A, 10X, 13X, AW-300, AW-500 |
| | C | 1/16 in | 0.3 | 44 | 3-5 | 600- | Siliporite |
| | S | 1/8 in | | | | 700 | NK10, 20, 30 |
| | P | < 400 | – | 31 | 3-6 | 700 | Siliporite NK10AP, NK20AP |
| Magnesia-silica gel | G | V | 0.33 | 30 | – | 300 | Florisil |
| Fuller's earth | G | V | 0.54 | 40 | – | 130-250 | Cecacite Clarsil PCS-G |
| Silica gel | G | V | V | 27-45 | V | 300-800 | Davison Silica Gel |
| | G | V | 0.35-0.5 | 40-48 | 20-40 | 650-900 | Cecagel Sorbsil |
| | S P | 1/8 in | 0.34-0.51 | 41-52 | 21-28 | 650-700 | Cecagel Mobil Sorbead R, H |
| | S | 4-8 8 | 0.45 | 46-51 | 72 | 250 | Mobil Sorbead W |
| *–Carbons–* | | | | | | | |
| Shell-based | G | V | 0.5-0.6 | 27-34 | 20 | 800-1100 | Cochranex FCB |

*(continued)*

**Table 1.1.**    Physical Properties of Adsorbents *(continued)*

| Material | Shape | Size (Mesh) | $\epsilon_p$ (–) | $\rho_b$ (lb/ft³) | $r_p$ (Å) | $A_s$ (m²/g) | Commercial Products |
|---|---|---|---|---|---|---|---|
| Shell-based | | | | | | | Pittsburgh PCB |
| | G | V | 0.5-0.8 | 25-35 | 20-30 | 1000-1600 | Picactif T.A., T.E. |
| | G P | V | 0.6-0.65 | 27-36 | 18-19 | 1200-1500 | Acticarbone NC, WNC |
| | G | V | 0.5 | 27-32 | 20 | 800-1100 | Barnebey-Cheney AC, KE, PC, VG |
| | G | 4-6 8-30 | 0.5 | 33-35 | 20 | 800-900 | Girdler 32E, 32W |
| | P | – | 0.6-0.8 | 20-22 | 30 | 1200 | Barnebey-Cheney JF, JU, YF |
| Wood-based | C P | 1/2 in 1/3 in | 0.7-0.75 | 12-28 | 22-24 | 750-1450 | Acticarbone AC Anticromos |
| | P | V | 0.3-0.5 | 9-35 | – | 600-1200 | Darco KB, G60 Nuchar Aqua Nuchar WA, B-100, C, C-100, C-115, C-190, CEE |
| | G | 5-7 | – | 24 | 5-10 | 1400 | Supersorbon W |

*(continued)*

Table 1.1. Physical Properties of Adsorbents *(continued)*

| Material | Shape | Size (Mesh) | $\epsilon_p$ (–) | $\rho_b$ (lb/ft³) | $r_p$ (Å) | $A_s$ (m²/g) | Commercial Products |
|---|---|---|---|---|---|---|---|
| Wood-based | P | V | – | 21-27 | 8-30 | 1000-1500 | Carboraffin |
| | P | V | – | 27-29 | 3-10 | 750-900 | Brilonit |
| | G | 10-30 | 0.6 | 20 | 20-40 | 600-1000 | Cochranex FCA, FCC, FCN-1, -2 |
| | G | V | 0.4-0.5 | 15-20 | 20-100 | 800-1200 | Picactif C, O |
| | P | 100-300 | 0.4-0.6 | 27-33 | – | 600-1200 | Picactif CM |
| Peat-based | C, G P | V | 0.55 | 15-32 | 30-40 | 500-1600 | Norit |
| | G | 5-7 | – | 20-24 | 5-20 | 1300-1400 | Supersorbon |
| | P | 200 | – | 28-31 | – | 700-900 | Acticarbone AM, AH |
| Coal-based | G | V | 0.65 0.75 | 20-30 | 20-38 | 500-1200 | Darco Granular Permutit Carbo-Dur |
| | G | 12-40 | 0.6 | 30 | 60-65 | 800-1000 | Cochranex FCP-1, -2, FCW-V |
| | G P | V | 0.56-0.67 | 25-30 | 20 | 1000-1400 | Pittsburgh CAL F400, BPL, SGL |
| | G | – | 0.8 | 28 | 22 | 110 | Barnebey-Cheney MN |

*(continued)*

**Table 1.1.** Physical Properties of Adsorbents *(continued)*

| Material | Shape | Size (Mesh) | $\epsilon_p$ (–) | $\rho_b$ (lb/ft$^3$) | $r_p$ (Å) | $A_s$ (m$^2$/g) | Commercial Products |
|---|---|---|---|---|---|---|---|
| Coal-based | G | V | – | 27-37 | – | 850-1350 | Acticarbone LM Nuchar WV-G, -H, WV-L, -W |
| | G | 5-7 | – | 20-24 | 5-15 | 1300-1500 | Contarbon |
| | P | – | – | 25-30 | – | 600-700 | Darco BG, DC Hydrodarco B |
| Petroleum-based | C P | V | 0.65-0.85 | 30 | 18-22 | 800-1100 | Columbia |
| | S | 20-30 | 0.265 | 25-35 | 32 | 800-1100 | Union Carbide BAC |
| *–Organics–* | | | | | | | |
| Porous resin | S | 16-50 | – | 20-45 | – | 3 | Asmit 224 Duolite S-35 |
| | G | | | | | | Permutit S-360 Wofatit E |
| | S P | 16-50 | – | 30-40 | – | – | Ionex RV |
| Cross-linked | G | 50-140 | – | 19-22 | 500 | 15-35 | Chromosorb 101, 103 |

*(continued)*

Table 1.1.    Physical Properties of Adsorbents *(continued)*

| Material | Shape | Size (Mesh) | $\epsilon_p$ (-) | $\rho_b$ (lb/ft$^3$) | $r_p$ (Å) | $A_s$ (m$^2$/g) | Commercial Products |
|---|---|---|---|---|---|---|---|
| poly-styrene | S | 20-50 | 0.4-0.45 | 40-44 | 90 | 330 | Amberlite XAD-2 |
|  | S | 20-50 | 0.5-0.55 | 39 | 50 | 750 | Amberlite XAD-4 |
|  | G | 60-200 | – | 18-20 | 85 | 300-400 | Chromosorb 102 |
| Phenolic | G | 16-50 | – | 22-25 | – | – | Duolite S-30, -37, ES-33 |
| Acrylic ester | S | 20-50 | 0.5-0.55 | 41 | 80 | 450 | Amberlite XAD-7 |
|  | S | 20-60 | 0.5-0.54 | 43 | 250 | 140 | Amberlite XAD-8 |
| Aromatic-amine resin | G | 10-50 | – – | 40-50 | – | – | Asmit 173N |
| Quaternary amine chloride resin | S | 16-50 | 0.65 | 40-50 | – – | – – | Asmit 259N, 261 Duolite ES-111, A-140 |
| Copper-amine resin | S | 10-50 | – | 30 | – | – | Duolite S-10 |
| Cellulose | G | 100-200 | – | – | – | 10000 | Whatman CC31, CF1, CF2, 11, 12 |

$\epsilon_p$ = particle porosity; $\rho_b$ = bulk density; $r_p$ = pore size; $A_s$ = surface area.

C = cylindrical pellets; G = granular; P = powder; S = spherical beads; T = tablets; V = various.

vated sludge treatment methods. Furthermore, the process is suitable for the control of certain inorganic compounds such as heavy metals, reduced sulfur gases, and chlorine.

## PROCESS CONFIGURATION

The simplest adsorption unit configuration is the mixed batch reactor (Figure 1.1a). In this case, the adsorbent is mixed with the polluted stream and the system is allowed to approach equilibrium. At the end, the two phases are separated for subsequent treatment or disposal. Powdered activated carbon is almost exclusively used as the adsorbate in this configuration. Unless the equilibrium conditions are extremely favorable, the use of a single batch reactor results in low removal efficiency and requires the use of pure carbon. This is because the desired high effluent quality equilibrates with a low solids concentration.

The use of two or more batch reactors in series, as shown in Figure 1.1b, improves the efficiency of the process. The spent adsorbent from the first reactor leaves the process, while the used adsorbent from any stage is sent to the preceding one. The result of this countercurrent movement is that all the discarded adsorbent is in equilibrium with the high concentration of the first reactor, and thus heavily loaded. At the same time, the effluent from the last stage can be of a high quality. As the number of stages increases, the efficiency increases. The complexity of the process also increases however, and thus simplicity, the main advantage of a batch configuration, is lost. Additionally, the time required to approach equilibrium at low solute concentrations can be extremely long. Consequently, the last reactors may have to be of unreasonable size.

Adsorption in the batch mode is seldom used because it is highly inefficient compared to continuous plug flow configurations. However, in some cases it is the only option available. For example, an extremely high level of suspended solids or viscosity in feed solution may render a columnar process hydraulically impractical. The addition of powdered activated carbon to an activated sludge process to enhance the removal of organic compounds from industrial wastewater is a very common practice.

The most efficient arrangement for conducting adsorption operations is the columnar continuous plug flow configuration known as a fixed bed. In this adsorption mode, the reactor consists of a packed bed of adsorbent through which the stream under treatment is passed. As the polluted stream travels through this bed, adsorption of the con-

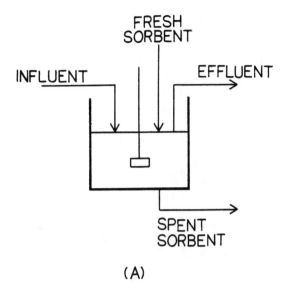

(A)

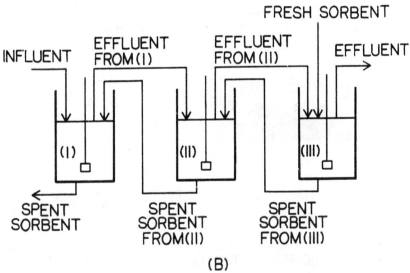

(B)

**Figure 1.1.**    Batch Configurations

taminants takes place and a purified effluent exits the column (Figure 1.2).

The portion of the bed close to the inlet is continuously contacted by the concentrated feed stream, whereas the subsequent portions are exposed to adsorbate not adsorbed by an earlier portion. Thus, the adsorbate becomes fully loaded at the inlet first, and then downstream.

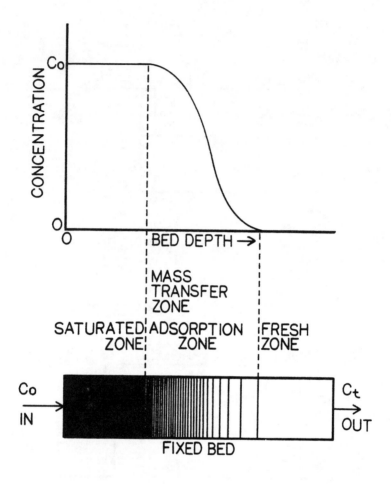

**Figure 1.2.**   Concentration Profile along Column

The part of the bed that displays a gradient in solid concentration from zero to equilibrium is called the mass transfer zone (MTZ). As the name indicates, this is the active part of the bed where adsorption actually takes place. The concentration of pollutants in the stream also changes continuously throughout this part of the bed, from a value close to zero at the beginning of the MTZ to the feed concentration at the end.

As the saturated part of the bed increases, the MTZ travels downstream and eventually exits the bed. This gives rise to a typical effluent concentration versus time profile, which is called the breakthrough curve (Figure 1.3). At breakthrough, the effluent concentration starts to rise and eventually reaches the effluent quality limit set by effluent

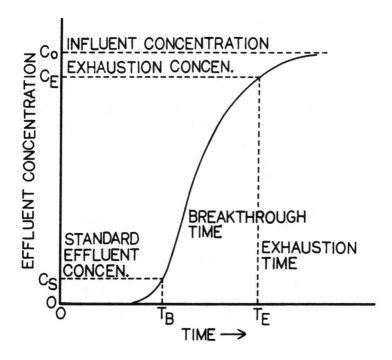

**Figure 1.3.**    Breakthrough Curve from a Fixed Bed Adsorber

regulations. This concentration is termed the "breakthrough concentration". The time it takes the effluent to reach this concentration is called "breakthrough time". Alternatively, this time can be described in terms of the amount of waste treated, expressed as the breakthrough volume or the breakthrough mass.

When the effluent concentration reaches its maximum, which is theoretically equal to that in the influent, the bed is considered to be exhausted. Exhaustion concentration, time volume, and mass are defined in a manner similar to those for breakthrough.

When the effluent quality becomes unacceptable, or after a predetermined time interval, the operation of the column is stopped and the spent adsorbent is either disposed of or regenerated. There are several regeneration methods applicable to various adsorption systems.

## REGENERATION

If activated carbon is used as an adsorbent and there are no metals adsorbed, thermal regeneration is always applicable. This method has several disadvantages: the spent carbon has to be first removed from the bed and sent to a specialized facility, the cost of which can be justified for only large installations. Furthermore, recovery or recycling of the adsorbed material is generally not possible by this method. However, thermal regeneration is widely practiced, since for many systems, it represents the only feasible option. This is particularly true for most of the carbon adsorption applications, in water and wastewater treatment. Wet oxidation and biological regeneration are two other regeneration methods that generally exclude the option of conservation of the adsorbed material. These methods usually find application in the regeneration of powdered activated carbon.

Regeneration of an adsorbent with simultaneous recovery of the adsorbate can be achieved if, after loading, the process is reversed by shifting the equilibrium point away from the solid phase. This is accomplished in air adsorption systems by increasing the temperature by the use of hot gas or steam. Steam regeneration can be applied in water systems too, provided all the adsorbates are volatile organic compounds.

In chemical regeneration, the adsorbate is concentrated in a chemical agent, for which the adsorbate has a greater affinity than for the adsorbent. Typical agents used are solvents for organic adsorbates, and basic or acidic solutions when the adsorbate is a weak acid or base, respectively. This agent, as well as any other regenerating stream (steam, hot gas, etc.), is called regenerant.

*PRACTICAL GUIDELINES*

When the fixed bed configuration is used, the simplest and most economically efficient way to apply the regenerant is inside the column, with the adsorbent bed in place. The flow direction of the regenerant can be cocurrent or countercurrent to the polluted stream.

The countercurrent option is advantageous because the vast amount of adsorbate extracts the heavy loaded adsorbate close to the inlet, leaving the column without passing through the lighter loaded region close to the outlet. This results in deeper regeneration of the adsorbent, more efficient regenerant utilization, and a more concentrated regeneration effluent favoring reuse.

The countercurrent system is almost exclusively used in columns treating air stream. In aqueous adsorption systems, however, the high cost of the flow distributors and collectors may offset the advantages of the countercurrent system. The cocurrent mode, using the piping system for both main stream and regenerant, is often preferred.

Another technical difference between air and water adsorption beds is the direction of the polluted stream. In air, the loading is generally downflow to prevent scouring of the bed. In contrast, upflow systems in water adsorption have kinetic advantages. This happens because any water or wastewater stream contains some suspended solids, which tend to be retained by the bed and foul it. Periodic backwashing, resulting in at least 50% expansion of the bed, is required to remove those solids. After backwashing, the bed is stratified with the larger particles on the bottom, and the smaller on the top. As will be discussed in later chapters, the rate of adsorption increases with increasing concentration. It is also inversely proportional to the square of the radius of the adsorbent particle. Thus, the top of a stratified bed is a much more efficient adsorbent than the bottom.

In a downflow loading, the combination of a low concentration with a less efficient adsorbent size, reduces the rate of adsorption at the bottom of the bed. This results in a higher effluent concentration and less efficient treatment. Yet, the downflow loading is often chosen because a careful hydraulic design is required for an upflow system, especially when a low density synthetic resin is used as the adsorbent.

Since regeneration or replacement of the adsorbent is necessary, continuous operation of an adsorption system requires the involvement of two or more beds. One of them undergoes regeneration while the rest contact the system.

If more than one bed is on-line, they can be arranged either in parallel or in series. In the first arrangement, the effluents of all the beds are blended. This allows each individual bed to be fully loaded until exhaustion, while the quality of the blended effluent still meets the standard.

When the beds are arranged in a series, the first bed, which has the heaviest load, is removed when the effluent from the last one is no longer acceptable. The second bed becomes the first bed and a fresh bed is placed at the end of the series (Figure 1.4).

Unless the MTZ is very short, those multiple bed configurations allowing complete exhaustion of the bed improve the utilization of the adsorbent. A single bed has to leave the process as breakthrough occurs, and its MTZ remains unsaturated.

*MOVING BED*

The ultimate evolution of the multiple bed in a series configuration is the moving bed shown schematically in Figure 1.5. In the system, the adsorbent is moved around a single loop of interconnected vessels, so that adsorption, regeneration, and backwashing are taking place

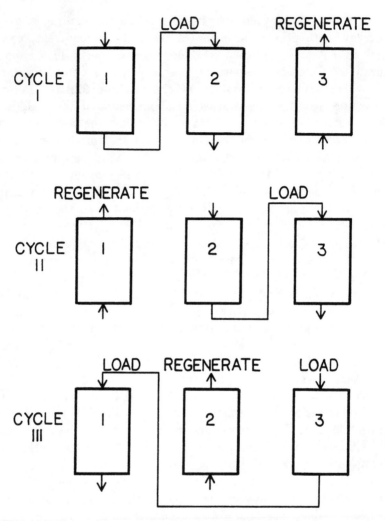

**Figure 1.4.**   Operation of a Three-Column Adsorption System

simultaneously at different locations. The efficiency is optimized when the system is operated countercurrently, with the adsorbate moving in one direction and all other streams in the opposite direction.

The moving bed offers several advantages. The process is truly continuous and this eliminates the need for standby equipment. Also, since all of the adsorbent is actively engaged in either adsorption, regeneration, or backwashing, minimum requirement of adsorbent inventory is achieved. Finally, the process reaches a steady-state condition, and effluent of constant concentration is received.

Due to the high degree of complexity and technology involved, the

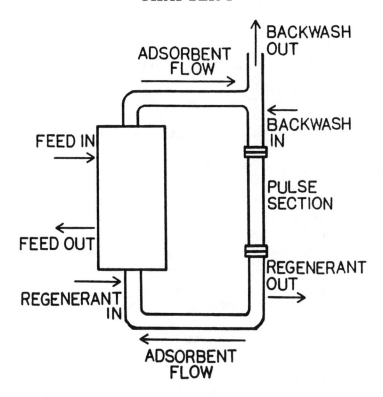

Figure 1.5.    Schematic Diagram of a Continuous Counter-Current Adsorption System

design of moving bed processes tends to be proprietary, and is economically justifiable for only large installations.

## SAFETY CONSIDERATIONS

There are some safety considerations when adsorption is applied. Since the process is exothermic, excessive heat may be accumulated in a bed treating a concentrated air stream. This heat may result in desorption and explosive ignition of the adsorbate. To eliminate this danger, the influent concentration is limited to one-fourth of the ignition limit of the adsorbate. Furthermore, the bed may be water-jacketed to permit convection of the heat generated. A similar problem exists in water systems employing polystyrene-based adsorbents exposed to nitric acid.

Activated carbon adsorbs oxygen to some extent, and it may eliminate it from the air in the free space of a column or a container. This can be fatal for workers who enter the vessel for maintenance;

breathing air must be added and maintained for safe entry.

## EXPERIMENTAL DESIGN AND TESTING

Adsorption is a complicated surface phenomenon, which has not been adequately studied. It is also an expensive process compared to other waste treatment methods. Therefore, extensive experimental investigations, at both the laboratory and pilot level, are essential for a successful application of this process.

During these tests, two questions are generally addressed: how much pollutant can be loaded onto an adsorbent, and how long will it take for this loading to occur. If non-thermal regeneration is considered, additional questions concerning the attainable degree of desorption, as well as the dosage of regenerant required, need to be investigated. Several alternative adsorbents and regenerants may have to be tested.

The answer to the first question comes from the determination of the equilibrium parameters. These are the most critical set of parameters, since they determine whether the process is feasible or not. Lab scale batch experiments are usually performed to study the equilibrium conditions. Methods and models for equilibrium studies are provided in subsequent chapters. The equilibrium experiments are simple and inexpensive and lead to a straightforward relationship between the concentrations of adsorbate in the solid and fluid phases.

Although the feasibility of the process depends on the equilibrium parameters, the system's efficiency is usually a strong function of the rate of adsorption. What the term "process efficiency" actually reflects in adsorption is the extent to which the adsorbent is loaded when it leaves the process for regeneration or disposal.

For many systems, adsorption may be a slow process requiring an unreasonably long time for its completion. Incomplete saturation of some of the adsorbent may, therefore, be inevitable. A kinetic study is thus required to optimize the utilization of the adsorbent within the technical and economical limits.

For batch configuration operating below saturation, the process can be easily simulated in the lab, using small batch reactors. The effect of residence time, number of stages, and dosage of adsorbent can thus be investigated. In a fixed bed, application of the kinetics process is much more complicated, and may require pilot scale experimentation. At breakthrough, the non-saturated part of an adsorption bed is its MTZ. It follows that the efficiency of the process is related to the size of the MTZ.

Indeed, if only one bed is used on line, its operation has to be stopped as soon as the breakthrough concentration is reached. Although the MTZ is not saturated yet, the entire bed has to be regenerated or replaced. Unless the MTZ is so short that it can be considered insignificant compared to total depth, an appreciable portion of the adsorbent is not completely used.

This waste of adsorbent can be reduced or eliminated by increasing the depth of the bed or using several beds in a series. In this latter case the first bed can become saturated and replaced, while the subsequent beds contain the MTZ. In both cases, increased total depth of bed means increases in equipment size, adsorbent inventory, heat loss, and their associated costs. In the worst case, the MTZ may be longer than the depth of an improperly design bed. In this case, breakthrough occurs immediately upon loading, and the unit may not be operative. It is obvious that the efficient design of a fixed bed adsorption process requires minimization and good knowledge of the size of MTZ. This is especially true for water adsorption systems, where the rate of adsorption tends to be slow and the MTZ long.

The shape and size of MTZ depends on many factors; the most important ones are the equilibrium parameters, the contact time, and the size of the adsorbent particle. While the first factor is beyond control of the designing engineer, the last two represent choices that have to be made.

The square of the diameter of the adsorbent particle is inversely proportional to the rate of adsorption, as well as to the frictional losses. Thus, the use of fine adsorbent particles will shorten the MTZ but increase the head loss. There is an optimum particle size that gives the best overall performance.

One way to evaluate the length of the MTZ, and choose a suitable depth of bed, is by conducting the following pilot test. Several beds are arranged in series, and allowed to contact the waste stream at a superficial velocity similar to that anticipated for the full scale column. The test runs until complete exhaustion of the last bed. The effluent concentration of each bed is measured and plotted versus time. A pilot of this kind is shown in Figure 1.6.

Systems consisting of a total depth equal to one, two, three and so on pilot beds are considered. The time it takes each system to reach breakthrough and exhaustion is plotted against the total depth of the system (Figure 1.7). The results provide two lines parallel to each other. The distance between those two lines, in a direction parallel to the axis of cumulative depth is equal to the length of the MTZ. Also, the same graph provides the relationship between bed depth

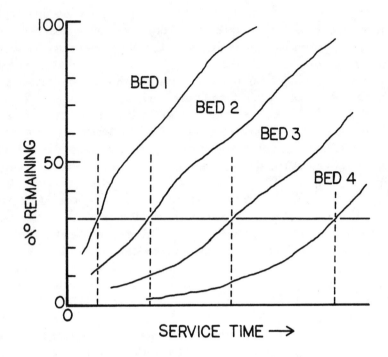

**Figure 1.6.**  Fixed Bed Series Breakthrough Curves

and service time. Given all this information, the optimum depth can be readily decided.

Alternatively, the effect of increasing contact time (or bed depth) on adsorption can be determined. This is done by plotting the ratio of adsorbent exhausted over waste treated, versus the contact time. For each of the cumulative bed depths, a single fixed bed exhaustion rate can be calculated. Figure 1.8 illustrates data treated in this manner, and shows carbon usage rate decreasing with contact time. The contact time, after which the adsorbent usage rate approaches a constant value, represents the optimum one.

To determine the degree of efficiency gained by using two or more beds in a series, an additional pilot test is performed. Two or more columns are set up in a series, each with a portion of the selected contact time. Column influent and effluent contaminant concentrations are monitored until the last column effluent reaches breakthrough. At this point, the first column is removed from service, the second receives the influent, and a fresh column is added at the final portion. The adsorbent usage rate achieved in this way is compared with that of a single bed configuration.

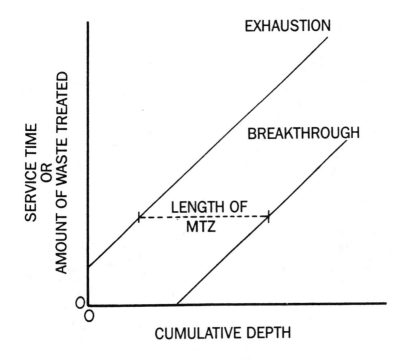

**Figure 1.7.**    Determination of MTZ Length

Although these pilot studies can lead to the secure design of an adsorption process, they suffer many disadvantages; they are costly and time-consuming. In the case of drinking water purification for example, a pilot test may run for several months. This inevitably limits the number of alternative solutions that can be tested, resulting in poor optimization of the process. Furthermore, the effect of future changes in the influent or in the plant can not be predicted.

The need and importance of developing a reliable, economical and faster method to predict fixed bed performance is obvious. The method that uses a spring balance, combining the mathematical advantage of a differential reactor and the simplicity of gravimetric measurement is an effective tool for the evaluation of adsorption parameters. Gravimetric measurements can be applied to column design, as indicated by the similarity of the rate parameters, measured with the differential reactor, to those from long column studies. The rest of this book is devoted to the presentation of promising developments in this direction.

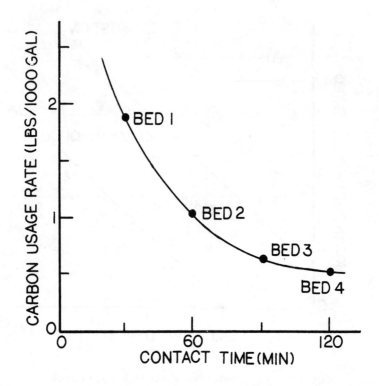

**Figure 1.8.** Effect of Increasing Contact Time of Fixed Bed
Carbon Usage Rate

# CHAPTER II

## ADSORPTION THEORY

## PRINCIPLES OF ADSORPTION

Adsorption is a surface phenomenon that can be defined as the increase in concentration of a particular component at the surface or interface between two phases. In any solid or liquid, atoms at the surface are subject to unbalanced forces of attraction normal to the surface plane. In discussing the fundamentals of adsorption, it is useful to distinguish between physical adsorption, involving only relatively weak intermolecular forces, and chemisorption, which involves essentially the formation of a chemical bond between the sorbate molecule and the surface of the adsorbent. Although this distinction is conceptually useful, many cases are intermediate and it is not always possible to categorize a particular system unequivocally [1].

Physical adsorption can be distinguished from chemisorption according to one or more of the following criteria:

1. Physical adsorption does not involve the sharing or transfer of electrons and thus always maintains the individuality of interacting species. The interactions are fully reversible, enabling desorption to occur at the same temperature, although the process may be slow because of diffusion effects. Chemisorption involves chemical bonding and is irreversible.

2. Physical adsorption is not site specific; the adsorbed molecules are free to cover the entire surface. This enables surface area measurements of solid adsorbents. In contrast, chemisorption is site specific; chemisorbed molecules are fixed at specific sites.

3. The heat of physical adsorption is low compared to that of chemisorption; however, heat of adsorption is not usually a definite criterion. The upper limit for physical adsorption may be higher than 20 kcal/mol for adsorption on adsorbents with very narrow pores. The heat of chemisorption ranges from over 100 kcal/mol to less than 20 kcal/mol. Therefore, only very high or very low heats of adsorption can be used as a criterion for this type of adsorption process [2].

Chemisorption is characterized mainly by large interaction potentials that lead to high heats of adsorption that approach the value of chemical bonds. This fact, coupled with other spectroscopic, electron spin resonance, and magnetic susceptibility measurements, confirms that chemisorption involves the transfer of electrons and the formation of true chemical bonding between the adsorbate and the solid surface [3]. Because chemisorption involves chemical bonding, it often occurs at high temperatures and is usually associated with activation energy. Also, the adsorbed molecules are localized on specific sites and, therefore, are not free to migrate about the surface.

## ADSORPTION: INTERACTION FORCES

### Dispersion Forces

Physical adsorption on nonpolar solids is attributed to forces of interactions between the solid surface and adsorbate molecules that are similar to the van der Waals forces (attraction-repulsion) between molecules. The attractive forces that involve the electrons and nuclei of the system are electrostatic in origin and are termed dispersion forces. These forces exist in all types of matter and always act as an attractive force between adjacent atoms and molecules no matter how dissimilar. They are always present regardless of the nature of other interactions and often account for the major part of the adsorbate-adsorbent potential [3,4]. The nature of the dispersion forces was first recognized in the 1930s by London [5]. Using quantum mechanical calculations, he postulated that the electron motion in an atom or molecule would lead to a rapidly oscillating dipole moment. At any instant, the lack of symmetry of the electron distribution about the nuclei imparts a transient dipole moment to an atom or molecule that would average zero over a longer time interval. When in close proximity to a solid surface, each instantaneous dipole of an approaching molecule induces an appropriately oriented dipole moment in a surface molecule. These moments interact to produce an instantaneous attraction. These forces are known as dispersion forces because of their relationship, noted by London [5], to optical dispersion. The dipole-dispersion interaction energy can be determined by:

$$E_D = -\frac{C}{d^6} \qquad (2.1)$$

where $E_D$ = dispersion energy or potential, $C$ = a constant, and $d$ = distance of separation between the interacting molecules.

In addition to dipole-dipole interactions, other possible dispersion interactions contributing to physical adsorption include dipole-quadrapole and quadrapole-quadrapole interactions. If these two are included, the total dispersion energy [3,5] becomes:

$$E_D = -\frac{C}{d^6} - \frac{C'}{d^8} - \frac{C''}{d^{10}} \qquad (2.2)$$

where $C'$ = a constant for dipole-quadrapole intractions and $C''$ = a constant for quadrapole-quadrapole interactions.

The contribution to $E_D$ from the terms in Equation (2.2) clearly depends on the separation, $d$, between the molecules; therefore, the dipole-dipole interactions will be most significant. Quadrapole interactions involve symmetrical molecules with atoms of different electronegativities like $CO_2$. This molecule has no dipole moment but does have a quadrapole $(^-O - {}^+C^+ - O^-)$ that can lead to interactions with polar surfaces.

When an adsorbate molecule comes very close to a solid surface molecule to allow interpenetration of the electron clouds, a repulsive interaction will arise, which is represented semiempirically by the expression:

$$E_R = \frac{B}{d^{12}} \qquad (2.3)$$

where $E_R$ = repulsion energy and $B$ = a constant. The total potential energy of van der Waals interactions is the sum of the attractive energy and the repulsion energy:

$$E = -\frac{C}{d^6} + \frac{B}{d^{12}} \qquad (2.4)$$

The inverse sixth energy term falls rapidly with increasing $d$, but not nearly as rapidly as the repulsion term. Thus, the dispersion energy is more important than the repulsion at longer distances.

*Surface Tension*

The surface of a liquid in contact with its vapor has different properties from those of the bulk phase. A molecule in the interior of a liquid is surrounded on all sides by neighboring molecules of the same substance and, therefore, is attracted equally in all directions. A molecule at the surface, however, is subject to a net attraction toward the bulk of the liquid, in a direction normal to the surface, because the number of molecules per unit volume is greater in the bulk of the liquid than in the vapor. Because of the unbalanced attraction, the surface of a liquid always tends to contract to the smallest possible area. To extend the area of the surface, work must be done to bring the molecules from the bulk of the liquid into the surface against the inward attractive force. The surface portion of a liquid, therefore, has a higher free energy than the bulk liquid.

The work required to increase the area by 1 cm$^2$ is called the surface free energy. As a result of the tendency to contract, a surface behaves as if it were in a state of tension, and it is possible to ascribe a definite value to this surface tension, $\gamma$, which is defined as the force in dynes acting at right angles to any line of cm length in the surface. The work done in extending the area of a surface by a cm$^2$ is equal to the surface tension, which is the force per centimeter opposing the increase, multiplied by 1 cm, the distance through which the point of application of the force is moved. It follows, therefore, that the surface energy, in ergs per square centimeter, is numerically equal to the surface tension in dynes per centimeter. In more general terms, the work, $W$, done by the surface in extending its area, $A$, by an amount, $dA$, is:

$$dW = -\gamma dA = -dG \quad (2.5)$$

hence

$$dG = \gamma dA \quad (2.6)$$

where $dG$ is the change in free energy.

Since the surface energy is a Gibbs free energy, the surface enthalpy, $\Delta H$, can be evaluated from change of surface tension $\gamma$ with temperature. From the Gibbs-Helmholtz equation:

$$\Delta G = \Delta H - T(\frac{\partial G}{\partial T})_p \quad (2.7)$$

The surface enthalpy is given by:

$$\Delta H = \gamma - T(\frac{\partial \gamma}{\partial T})_p \quad (2.8)$$

For water at 20°C, $\gamma = 72.75$ erg/cm$^2$, $(\partial \gamma / \partial T)_p = -0.148$ (the negative sign is due to the decrease of $\gamma$ with increase in temperature), and so $\Delta H = 116.2$ erg/cm$^2$. This is the decrease in enthalpy with the destruction of 1 cm$^2$ of liquid surface.

The addition of a solute to a liquid may alter the surface tension considerably. In the case of aqueous solutions, solutes that can markedly lower the surface tension of water or organic compounds that contain both a polar hydrophilic group and a nonpolar hydrophobic group include, for example, organic acids, alcohols, esters, ethers, ketones, and so on. The hydrophilic group makes the molecule reasonably soluble while the hydrocarbon residues have low affinity for water and little work is required to bring them from the interior to the surface. Solutes that lower the surface tension tend to accumulate preferentially at the surface, and hence, there will be a greater proportion of the solute at the interface than the bulk of the solution. This represents a case of adsorption of the solute at the surface of the solution, and the solute is said to be positively adsorbed at the interface.

Electrolytes, salts of organic acids, bases of low molecular weight, and certain nonvolatile electrolytes usually increase the surface tension of aqueous solutions above the value for pure water. These increases are much smaller than the decreases produced by organic acids and similar compounds. The observed increases are attributed to ion-dipole interactions of the dissolved ions that tend to pull the water molecules into the interior of the solution. Additional work must be done against the electrostatic forces to create a new surface. The surface layers in such solutions have lower concentrations of the solute than in the bulk solution. The solute is said to be negatively adsorbed at the interface [6].

*Porosity*

Figure 2.1 shows the structures of graphite (a) and turbostratic carbon (b) [7]. Activated carbons are similar to the second type that have microcrystallites only a few layers thick and less than 100Å in width. The many structural imperfections in actived carbon's microcrystallites result in many possibilities for the edge carbons to react with their surroundings.

However, the porous nature of carbon determines its adsorptive properties. It has been stated that "over 99% of the active sites for adsorption in GAC are located in the interior of the particle" [8]. Figure 2.2 is a schematic of a constituent region of an activated carbon particle where the macropores have diameters of 30 to 10,000Å and the micropore diameters are in the 10 to 30Å range [9].

Tables 2.1 and 2.2 summarized the mercury and nitrogen pure volume data for several commercial and experimental carbons (12×40 U.S. mesh) [8]. It is obvious that pore surface area does indeed approach 99% of the total area. Also, a major part of the total $N_2$-BET surface area is found in the smallest micropores. In the seventh and eighth

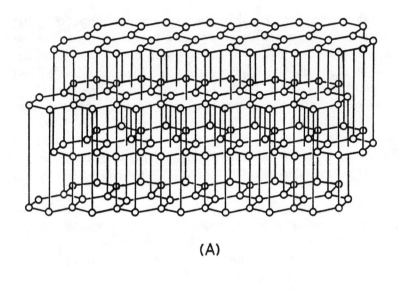

(A)

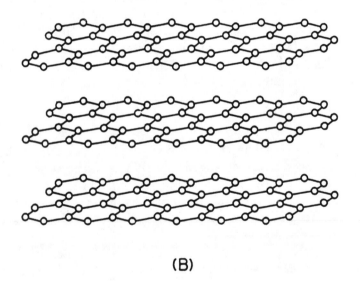

(B)

**Figure 2.1.**   Schematic Diagram Comparing (a), a Three-Dimensional Graphite Lattice, with (b), a Turbostratic Structure

columns, the effect of additional activation is noted.  GC-25B was steam activated for an hour longer than GC-25A+26. This increased the total $N_2$-BET surface area by 83% and opened some of the narrow pores to form wider pores.

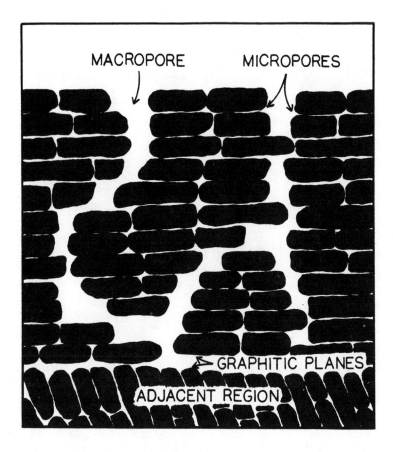

**Figure 2.2.** Schematic Representation of a Constituent Region of an Activated Carbon Particle

*Surface Functional Groups*

It has been shown that graphite, carbon black, and activated carbon irreversibly chemisorb molecular $O_2$ [10,11]. This chemically bound $O_2$ is removable only at elevated temperatures. In early research, study of the chemistry of these carbon surface oxides involved, indirect methods such as acid-base neutralization and thermal removal of CO and $CO_2$. More recent studies employed direct investigation of surface functional groups with infrared and ESR (electron spin resonance) spectroscopy [12].

In 1929, several observations were made with sugar-based carbons that were activated at high temperatures in comparison with carbons activated at low temperatures [13]. The former had markedly different

Table 2.1. Pore Volumes of Some Carbons

| Carbon | Pore Volume ($cm^3/g$) | | |
| | Hg | $N_2$ | Total |
|---|---|---|---|
| Commercial F | 0.500 | – | – |
| Commercial C | – | 0.24 | 0.60* |
| Commercial A | 0.556 | 0.69 | 0.99 |
| | | | 0.94* |
| Commercial D | – | 0.65 | 1.00* |
| Commercial B | – | 0.70* | – |
| GC-16D (1 hr) | 0.227 | – | – |
| GC-20G | 0.277 | 0.44 | 0.66 |
| GC-21 | 0.446 | – | – |
| GC-22 | 0.235 | 0.33 | 0.48 |
| GC-25A+26 (1 hr) | 0.465 | 0.48 | 0.78 |
| GC-25B (2 hr) | 0.958 | 0.90 | 1.29 |
| GC-27A | 0.885 | 0.95 | 1.43 |
| GC-27B | 0.984 | 0.99 | 1.61 |

* From the manufacturer's literature.

surface properties with respect to acid-base properties, electrophoretic mobility, and degree of hydrophilicity. The activation of C at 1,000°C in pure $CO_2$ or under a vacuum followed by exposure to $O_2$ at room temperature produces a carbon surface that is capable of raising the pH value of either a neutral or an acidic solution, is hydrophobic, and has a positive electrophoretic mobility [14]. In contrast, the oxidation of C by $O_2$ between 200 and 400°C or by adding C to an aqueous oxidizing solution yields a surface characteristic opposite to the previous.

A method of characterizing carbons that are activated and oxidized at different temperatures was developed from their acid-base properties [13]. Those low-temperature oxidized carbons that adsorb $OH^-$ ions primarily are called L-carbons. Those that are activated at high temperatures and adsorb $H^+$ ions are called H-carbons. These classifications have been found to divide above and below an activation-oxidation temperature of 500 to 600°C [12].

Almost every type of functional group known in organic chemistry has been suggested to be present on activated carbon's surface [12].

Table 2.2.    Surface Area versus Pore Radius and Total $N_2$ BET Surface Area

| Pore Radius (Å) | Com A | Com E | $N_2$ BET Surface Area (%) | | | | | |
| --- | --- | --- | --- | --- | --- | --- | --- | --- |
| | | | GC-18F | GC-20G | GC-21 | GC-25A+26 | GC-25B | GC-27A |
| 300-250 | 0.0 | 0.1 | | 0.0 | 0.0 | | 0.0 | 0.0 |
| 250-200 | 0.1 | 0.0 | | 0.1 | 0.0 | | 0.0 | 0.0 |
| 200-150 | 0.1 | 0.1 | | 0.1 | 0.1 | | 0.1 | 0.1 |
| 150-100 | 0.2 | 0.1 | | 0.0 | 0.0 | | 0.1 | 0.2 |
| 100-90 | 0.1 | 0.1 | | 0.0 | 0.0 | 0.0 | 0.0 | 0.1 |
| 90-80 | 0.1 | 0.1 | | 0.0 | 0.0 | | 0.1 | 0.2 |
| 80-70 | 0.1 | 0.1 | 0.0 | 0.0 | 0.0 | | 0.1 | 0.3 |
| 70-60 | 0.2 | 0.2 | | 0.0 | 0.0 | | 0.1 | 0.5 |
| 60-50 | 0.2 | 0.1 | | 0.1 | 0.1 | 0.1 | 0.1 | 0.5 |
| 50-40 | 0.4 | 0.2 | | 0.1 | 0.0 | 0.0 | 0.2 | 1.0 |
| 40-30 | 0.7 | 0.3 | | 0.0 | 0.0 | 0.0 | 0.3 | 1.5 |
| 30-20 | 2.1 | 0.9 | 0.1 | 0.2 | 0.4 | 0.3 | 0.8 | 2.1 |
| 20-15 | 3.2 | 2.1 | 0.4 | 0.5 | 0.9 | 0.8 | 1.4 | 2.2 |
| 15-20 | 13.3 | 15.0 | 3.0 | 4.3 | 5.4 | 4.3 | 13.3 | 10.7 |
| 10-7 | 79.1 | 80.7 | 96.4 | 94.4 | 92.9 | 94.2 | 83.4 | 80.9 |
| Total $N_2$ BET $A_s$ (m²/g) | 1163.4 | 404.3 | 1013.5 | 761.0 | 1190.2 | 853.6 | 1560.6 | 1377.0 |

$A_s$ =Surface area

The ones suggested most often are carboxyl groups (I), phenolic hydroxyl groups (II), and quinone-type carbonyl groups (III). Other suggested groups are ether, peroxide, and ester groups in the forms of normal (IV) and fluorescein-like lactones (V), carboxylic acid anhydrides (VI), and the cyclic peroxide (VII).

*The Polanyi Theory*

Manes [15,16] has presented the Polanyi theory in an effort to de-

(I)    (II)    (III)

(IV)    (V)    (VI)    (VII)

scribe adsorption phenomenon. To this end, quantification of adsorption potential is required and it is defined as the amount of work required in moving a molecule from an adsorbent's pore to infinity. The adsorption space (pore volume) is assumed to be composed of many equipotential adsorption energy surfaces which vary from a maximum in the finest pores to zero in the bulk solution. Hence, for a given adsorbent, the equipotential surface should be the same for all adsorbates. Consequently, a plot of adsorbate volume versus adsorption potential should result in a single characteristic curve good for all adsorbates on that particular adsorbent. This theory then suggests that an adsorption isotherm could be predicted for any compound or any adsorbent for which a characteristic adsorption curve has been obtained.

*Isostere and Isobar*

Two additional adsorption equilibrium relationships are the isostere and the isobar. The isostere is a plot of the $\ln p$ versus $1/T$ at a constant amount of vapor adsorbed. Adsorption isostere lines are usually straight for most adsorbate-adsorbent systems. Figure 2.3 is an adsorption isostere graph for the adsorption of $H_2S$ gas onto molecular sieves. The isostere is important in that the slope of the isostere corresponds to the heat of adsorption. The other equilibrium relationship is the isobar. The isobar is a plot of the amount of vapors

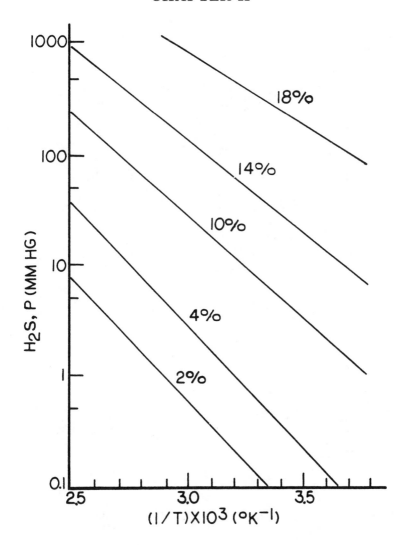

**Figure 2.3.**    Adsorption Isosteres of $H_2S$ on 13X Molecular Sieve
Loading in % Weight

adsorbed versus temperature at constant pressure.

Figure 2.4 shows several isobar curves for the adsorption of ammonia vapors on charcoal. Note that the amount adsorbed decreases with increasing temperature, which is always the case for physical adsorption.

Since isotherm, isostere, and isobar were developed at equilibrium conditions, they depend on each other. By determining one, such as the isotherm, the other two relationships can be determined for a given

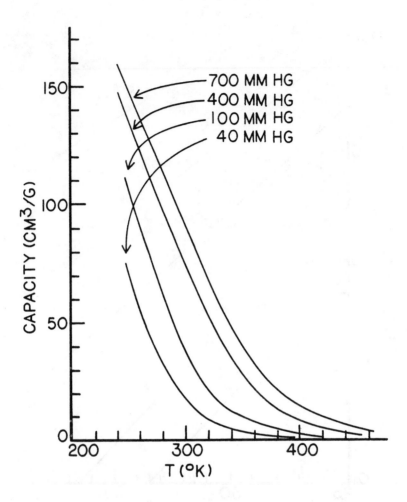

**Figure 2.4.**    Isobars for Adsorption of Ammonia on Charcoal

system. In the design of a pollution control system, the adsorption isotherm is by far the most commonly used equilibrium relationship.

*Classification of Adsorption Isotherms*

The majority of physiosorption isotherms may be grouped into six types shown in Figure 2.5 [17]. In most cases at sufficiently low surface coverage, the isotherm reduces to a linear form (i.e. $n^a = P$), which is often referred to as the Henry's Law region. On heterogeneous surfaces, this linear region may fall below the lowest experimentally measurable pressure.

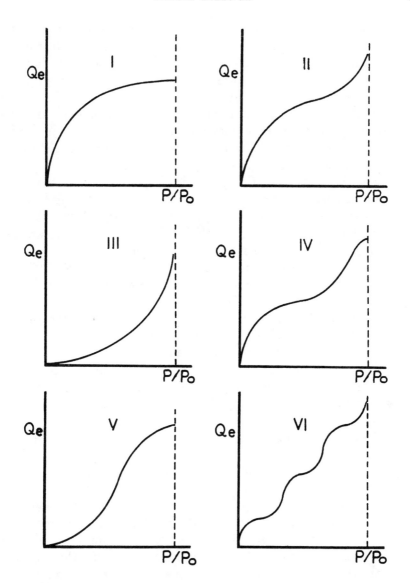

**Figure 2.5.**    Types of Isotherms for Gaseous Adsorption

The reversible Type I isotherm (or Langmuir isotherm) is concave to the $P/P_0$ axis and $n$ approaches a limiting value as $P/P_0$ approaches 1.0. Type I isotherms are given by microporous solids having relatively small external surfaces (e.g., activated carbon, molecular sieves), the limiting uptake being governed by the accessible micropore volume rather than by the internal surface area.

The reversible Type II isotherm is the normal form of isotherm obtained with a nonporous or microporous adsorbent. The Type II isotherm represents unrestricted monolayer-multilayer adsorption. The beginning of the almost linear middle section of isotherm is often taken to indicate the stage at which monolayer coverage is complete and multilayer adsorption about to begin.

The reversible Type III isotherm is convex to the $P/P_0$ axis over its entire range and therefore does not exhibit any inflection points. Isotherms of this type are not common; the best known are found with water vapor adsorptions on pure nonporous carbons.

Characteristic features of Type IV isotherms are their hysteresis loops, which are associated with capillary condensation taking place in mesopores (pore size between micropore and macropore), and the limiting uptake over a range of high $P/P_0$. Type IV isotherms are obtained for many mesoporous industrial adsorbents.

The Type V isotherm is uncommon; it is related to the Type III isotherm in that the adsorbent-adsorbate interaction is weak, but is obtained with certain porous adsorbents.

The Type VI isotherm represents stepwise multilayer adsorption on a uniform nonporous surface. The step height represents the monolayer capacity for each adsorbed layer and, in the simplest case, remains nearly constant for two or more adsorbed layers. Examples of these kinds of isotherms are obtained with argon or krypton on graphitized carbon blacks at liquid nitrogen temperatures.

## EQUILIBRIUM MODELS

### Single-Component Adsorption

In order to successfully represent the dynamic adsorptive behavior of any substance from the fluid to the solid phase, it is important to have a satisfactory description of the equilibrium state between the two phases composing the adsorption system. In the following paragraphs, some of the more well-known single component adsorption equilibrium models are reviewed.

Henry's Law. The simplest adsorption isotherm, in which the amount adsorbed varies directly with the equilibrium gas, is often referred to as Henry's law:

$$q = KC \tag{2.9}$$

where $K$ = isotherm constant, $C$ = equilibrium concentration, and $q$ = equilibrium adsorbed amount on the adsorbent.

Linear isotherms have been found experimentally for the adsorption of some gas-solid systems [18-20]. In these systems the adsorbed layer is extremely dilute, the amount adsorbed corresponding to a fraction of one percent of the monolayer capacity.

Freundlich Isotherm. The equilibrium relationships in adsorbers can often be described by a Freundlich relationship [21], provided there is:
1. No association or dissociation of the molecules after they are adsorbed on the surface.
2. A complete absence of chemisorption.

In other words, for Freundlich isotherms to be valid, the adsorption must be purely a physical process with no change in the configuration of the molecules in the adsorbed state. Freundlich proposed the equation:

$$q = KC^{1/n} \qquad (2.10)$$

where $K,n$ = empirical constants dependent on the nature of solid and adsorbate, and on the temperature.

The Freundlich isotherm equation may be derived by assuming a heterogeneous surface with adsorption on each class of sites [22] that obey the Langmuir equation. According to the Freundlich equation, the amount adsorbed increases infinitely with increasing concentration or pressure. This equation is, therefore, unsatisfactory for high coverages. At low concentration, this equation does not reduce to the linear isotherm. In general, a large number of the experimental results in the field of van der Waals adsorption can be expressed by means of the Freundlich equation in the middle concentration range.

Langmuir Isotherm. The Langmuir model was originally developed to represent chemisorption on a set of distinct localized adsorption sites. The derivation of the Langmuir adsorption isotherm involves five implicit assumptions [23]:
1. The adsorbed gas behaves ideally in the vapor phase.
2. The adsorption gas is confined to a monomolecular layer.
3. The surface is homogeneous, that is, the affinity of each binding site for gas molecules is the same.
4. There is no lateral interaction between adsorbate molecules.
5. The adsorbed gas molecules are localized, that is, they do not move around on the surface.

The commonly quoted form is:

$$q = \frac{Q_0 bC}{1 + bC} \qquad (2.11)$$

where $Q_0$ = a temperature-independent constant which is supposed to represent a fixed number of surface sites, $b$ = a temperature dependent equilibrium constant.

This expression shows the correct asymptotic behavior for monolayer adsorption since at saturation $C \to \infty$, $q \to Q_0$, while at low sorbate concentration Henry's law is approached.

A simple kinetic derivation or a more refined thermodynamic derivation shows that the temperature-dependent equilibrium constant, $b$, follows a vant Hoff equation [24]:

$$b = b_0 \exp(-\Delta H / RT) \qquad (2.12)$$

$$b_0 = \frac{N_{av}\sigma^o t_0}{(2M_A RT)^{1/2}} \qquad (2.13)$$

where $\Delta H$ = heat of adsorption, $b_0$ = the nature of a frequency factor, $\sigma^o$ = the area of a site, $t_0$ = a constant about $10^{-12} - 10^{-13}$ sec, and $N_{av}$ = Avogadros number.

Since adsorption is exothermic $(\Delta H < 0)$, $b$ should decrease with increasing temperature. Some workers [25-27] have successfully applied this equation to estimate the heat of adsorption.

BET Isotherm. In 1938, Brunauer, Emmett, and Teller showed how to extend Langmuir's approach to multilayer adsorption, and their equation has come to be known as the BET equation. The basic assumptions are that each molecule in the first adsorbed layer is considered to provide one site for the second and subsequent layers, that the molecules in the second and subsequent layers, which are in contact with other sorbate molecules rather than with the surface of the adsorbent are considered to behave essentially as the saturated liquid, while the equilibrium constant for the first layer of molecules in contact with the surface of the adsorbent is different. The resulting equation for the BET equilibrium isotherm is:

$$\frac{q}{Q_0} = \frac{bP/P_s}{(1 - P/P_s)(1 - P/P_s + bP/P_s)} \qquad (2.14)$$

where $P_s$ = the saturation vapor pressure of the saturated liquid sorbate at the relevant temperature.

This expression has been found to provide a good representation of experimental physical adsorption isotherms, provided that the range of reduced pressure is restricted to $0.35 < P/P_s < 0.5$.

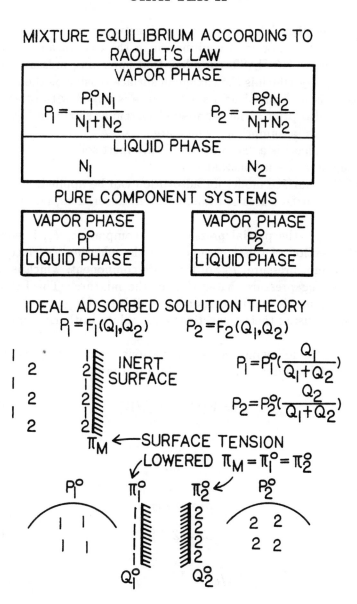

**Figure 2.6.** Predicting Solid-Gas Equilibria by Ideal Adsorbed Solution Theory and Its Analogy to Raoult's Law

*Multicomponent Adsorption*

Ideal Adsorbed Solution Theory. Myers and Prausnitz [28] proposed and successfully used an ideal adsorbed solution theory (IAST), to predict the adsorption of binary gas mixtures onto activated carbon.

Similar to the Raoult's law which describes vapor-liquid equilibrium, IAST describes vapor-solid equilibrium in the case of adsorption as shown in Figure 2.6. IAST considers that spreading pressure, an intensive property which is the change of surface tension as the result of adsorption, is unchanged as a single solute is mixed with other components in adsorption. The vapor-solid system as shown in Figure 2.6 includes several assumptions, which are:

1. The solid phase is a thermodynamically inert solid.
2. The temperature is constant.
3. The Gibbs surface model for adsorption can be applied.
4. The same surface area is available to all adsorbates, i.e., no steric effects.

Accordingly, the partial pressure of a component, $P_i$, is equal to the mole fraction of this component on the adsorbent surface times the equilibrium pressure, $P_i^o$, of this pure component, which is at the same spreading pressure, $\pi$, as that of the mixture. The IAST has been used to predict multicomponent equilibria of liquid-solid or gas-solid systems using their respective single-solute isotherm parameters [29-31].

In IAST, the following six basic equations are used to predict multicomponent behavior from single-solute isotherms:

$$q_i = V(C_{0i} - C_i)/W_s \qquad (2.15)$$

$$q_T = \sum_{i=1}^{N} q_i \qquad (2.16)$$

$$Z_i = q_i/q_T \qquad (2.17)$$

$$C_i = Z_i C_i^o \qquad (2.18)$$

$$1/q_T = \sum_{i=1}^{N} Z_i/q_i^o \qquad (2.19)$$

$$\frac{\pi_m A}{RT} = \int_0^{q_1^o} \frac{d\ln C_1^o}{d\ln q_1^o} dq_1^o = \frac{\pi_1^o A}{RT} = \cdots$$
$$= \int_0^{q_N^o} \frac{d\ln C_N^o}{d\ln q_N^o} dq_N^o = \frac{\pi_N^o A}{RT} \qquad (2.20)$$

where $V$ = volume of solution, $W_s$ = weight of adsorbent, $A$ = surface area of adsorbent, $T$ = temperature, $R$ = gas constant.

Equation (2.15) is the mass balance equation. Equation (2.16) sums the individual loadings, $q_i$, to get the total surface loading, $q_T$. Equation (2.17) is the definition of the mole fraction, $Z_i$, on the surface of the carbon for component $i$. Equation (2.18) is analogous to Raoult's law, where $C_i^o$ is the single-solute gas-phase surface concentration in equilibrium with $q_i^o$. The single-solute surface loadings, $q_i^o$, are the loadings that cause the same spreading pressures as are found in the mixture. Equation (2.19) is the expression for no area change per mole upon mixing in the mixture. Equation (2.20) equates the spreading pressures, $\pi_i^0$, of the pure component systems.

If the Freundlich isotherm, Equation (2.10), is used to represent single-solute behavior, a simpler relationship for predicting multicomponent adsorption can be derived [32] as follows:

$$\frac{d \ln C_i^o}{d \ln q_i^o} = n_i \qquad (2.21)$$

Accordingly, Equation (2.20) will be simplified to the expression:

$$n_1 q_1^o = \cdots = n_N q_N^o \qquad (2.22)$$

Combining those equations, the following equation for each adsorbate was derived:

$$C_i = \frac{q_i}{\sum_{j=1}^{N} q_j} \left[ \frac{\sum_{j=1}^{N} n_j q_j}{n_i K_i} \right]^{n_i} \qquad (2.23)$$

This equation has been used in the derivation of the equilibrium column model directly by applying the Newton-Raphson algorithm.

## ADSORPTION KINETICS

The adsorption on a solid takes place in several stages [33,34] as:
1. External diffusion: The mass transfer by diffusion of the adsorbate molecules from the bulk fluid phase through a stagnant boundary layer surrounding each adsorbent particle to the external surface of the solid.
2. Internal diffusion: Transfer of the adsorbate to the interior of the particle by the migration of the adsorbate molecules from the relative small external surface of the adsorbent to the surfaces of the pores within each particle and/or by the diffusion of the adsorbate molecules through the pores of the particles.
3. The actual adsorption process: The molecules in the pores are adsorbed from the solution to the solid phase. This stage is relatively

fast, compared to the first two steps; hence, local equilibrium is usually assumed between these two phases.

*External Diffusion*

The rate of transfer, $N_A$, for the first stage can be expressed by Equation (2.24):

$$N_A = \frac{k_f a_p \epsilon}{\rho_b}(C - C_s) \qquad (2.24)$$

where $k_f$ = external film mass transfer coefficient, $a_p$ = external surface area of the particle, $\epsilon$ = voidage between granules, $\rho_b$ = bulk density of the particle, and $C, C_s$ = concentration in the bulk solution and at the surface, respectively.

The external mass transfer coefficient for the fixed-bed adsorber can be estimated using the following equations:

1. Wakao and Funazkri [35]:
$$k_f = \frac{D_m}{2R_p}(2.0 + 1.1 Re^{0.6} Sc^{0.33}) \qquad (2.25)$$

2. Petrovic and Thodos [36]:
$$k_f = 0.355 V_s \left(\frac{1-\epsilon}{\epsilon}\right) Re^{-0.359} Sc^{-0.67} \qquad (2.26)$$

3. Lightfoot et al. [37]:
$$k_f = 2.23 V_s (1-\epsilon)^{1.51-0.49} Re^{-0.51} Sc^{-0.67} \qquad (2.27)$$

4. Chu et al. [38]:
$$k_f = 5.7 V_s Re^{-0.75} Sc^{-0.67} \qquad (2.28)$$

5. Ranz and Marshall [39]:
$$Sh = \frac{2k_f R_p}{D_m} = 2.0 + 0.6 Re^{0.5} Sc^{0.33} \qquad (2.29)$$

6. Wilke and Hougen [40]:
$$k_f = 1.82 V_s \left(\frac{1-\epsilon}{\epsilon}\right) Re^{-0.51} Sc^{-0.67} \qquad (2.30)$$

7. Williamson et al. [41]:

$$k_f = 2.40 V_s Re_p^{-0.66} Sc^{-0.58} \qquad Re = 0.08 - 125$$
$$= 0.442 V_s Re_p^{-0.31} Sc^{-0.58} \qquad Re = 125 - 5,000 \qquad (2.31)$$

where Reynolds number $Re = 2R_p V_s \rho_f / \mu$, $Re_p = 2R_p V_s \rho_f / \epsilon \mu$, Schmidt number $Sc = \mu / D_m \rho_f$, Sherwood number $Sh = 2k_f R_p / D_m$, $V_s$ = superficial velocity, $R_p$ = adsorbent pellet radius, $\mu$ = viscosity of fluid, $\rho_f$ = density of fluid, and $D_m$ = molecular diffusivity.

Among those correlations, Equations (2.25), (2.26), and (2.31) have been widely applied to packed beds. Generally, the values of the external mass transfer coefficient predicted by Equation (2.25) are considerably higher than the values obtained from Equations (2.26) and (2.31) because the axial dispersion effect was considered in the derivation. In the following study, the Petrovic and Thodos correlation for gas/solid systems, and the Williamson correlation for liquid/solid systems will be applied to fixed-bed breakthrough prediction.

*Internal Diffusion*

The rate of adsorption in porous adsorbents is generally controlled by transport within the pore network. It is convenient to consider intraparticle transport as a diffusive process and to correlate kinetic data in terms of a diffusivity defined in accordance with Fick's first law of diffusion:

$$N_A = D \frac{\partial C_p}{\partial r} \qquad (2.32)$$

where $D$ = intraparticle diffusivity, $C_p$ = concentration in the pore space.

Intraparticle diffusion may occur by several different mechanisms (ordinarily diffusion or molecular diffusion, Knudsen diffusion and surface diffusion), depending on the pore size, the sorbate concentration, and other conditions. If the pores are large and the solution relatively dense, the process is that of molecular diffusion. If the solution is a gas phase where density is low, or if the pores are quite small, or both, the molecules collide with the pore wall much more frequently than with each other. This process is known as Knudsen diffusion. Molecules adsorbed on a solid surface may evidence considerable mobility. Transport by movement of molecules over a surface is known as surface diffusion.

For a gas-solid system, the effective molecular diffusivity, $D_{m,eff}$, can be calculated using Chapman-Enskog equation [42] correlated with tortuosity, $\tau$, and particle porosity, $\epsilon_p$, as follows:

$$D_{m,eff} = \frac{\epsilon_p}{\tau} D_m \qquad (2.33)$$

$$D_m = \frac{BT^{3/2}(1/M_1 + 1/M_2)^{1/2}}{P\sigma_{12}^2 \Omega} \qquad (2.34)$$

$$B = (10.85 - 2.50\sqrt{1/(1/M_1 + 1/M_2)}) \times 10^{-4} \qquad (2.35)$$

where $D_m$ = molecular diffusivity, $M_1, M_2$ = molecular weight, $P$ = total pressure, $\sigma_{12} = (\sigma_1 + \sigma_2)/2$ collision diameter from the Lennard-Jones potential in Å, $\Omega$ = a function of $\epsilon/k_b T$, $\epsilon = \epsilon_1 \epsilon_2$ a Lennard-Jones force constant, $k_b$ = Boltzman constant.

For a liquid-solid system, the molecular diffusivity is usually estimated by the Wilke-Chang equation [43]:

$$D_m = 7.4 \times 10^{-8} \frac{(\phi M_b)^{1/2} T}{\mu V_a^{0.6}} \qquad (2.36)$$

where $\phi$ = an association factor of solvent (2.6 for water), $M_b$ = the molecular weight of solvent, and $V_a$ = molar volume of solute at its normal boiling temperature.

Kinetic theory provides the following relation for effective Knudsen diffusion in gases:

$$D_{k,eff} = 9700 \times r_p \sqrt{\frac{T}{M}} \times \frac{\epsilon_p}{\tau} \qquad (2.37)$$

where $r_p$ = pore radius.

Since there is a wide range of conditions under which both Knudsen and molecular diffusion are significant, a relation for use in this transition region was derived [44] as:

$$\frac{1}{D_{p,eff}} = \frac{1}{D_{k,eff}} + \frac{1}{D_{m,eff}} \qquad (2.38)$$

The tortuosity, a geometrical factor, should be independent of the diffusing gas and operating condition [45]. Wheeler [46] proposed a model in which the pores were visualized as cylinders of one fixed diameter, which interacts with any plane at an average angle of 45°, so that $\tau = 2$. Weisz and Schwartz [47] proposed a model in which $\tau = \sqrt{3}$. For diffusion through a randomly oriented system of long cylindrical pores, the tortuosity factor is 3.0 [48]. Another commonly used relation is $\tau = 1/\epsilon_p$ [49]. For accurate work $\tau$ must be determined experimentally.

## DIFFUSION MODELS

On the internal pore surface, an adsorbed molecule may hop along the surface when it attains sufficient activation energy and when an adjacent adsorption site is available. Although the mobility of the adsorbed phase will generally be smaller than that in the solution, the concentration is very much higher; so a significant contribution to the flux is possible. It has been shown that surface diffusion is very important when appreciable adsorption occurs [50,51]. Direct measurement of surface diffusion is not feasible, since the flux, due to diffusion through the solution is always present in parallel. In order to study surface diffusion, it is therefore necessary to eliminate the contribution from the fluid phase. The procedure which is illustrated in the study of Schneider and Smith [52], involves making measurements over a wide range of temperatures. The flux through the gas phase is determined from the high temperature measurements, since under these conditions the surface flux can be neglected. The flux through the gas phase at the lower temperature is then found by extrapolation and subtracted from the measured flux in order to estimate the surface flux.

### Surface Diffusion Model

This model pictures the diffusion of an adsorbate through an external film to the outer surface of a particle; at this surface, adsorption occurs instantaneously and equilibrium is assumed to be established between adsorbate in the fluid and that on the surface. The adsorbed material then diffuses into the pores in the adsorbed state. A material balance for spherical particle results in the following mass transfer equations:

$$\frac{\partial q}{\partial t} = \frac{1}{r^2}\frac{\partial}{\partial r}\left(r^2 D_s \frac{\partial q}{\partial r}\right) \tag{2.39}$$

$$k_f(C - C_s) = D_s \rho_p \frac{\partial q}{\partial r} \qquad \text{at } r = R_p \tag{2.40}$$

$$\frac{\partial q}{\partial r} = 0 \qquad \text{at } r = 0 \tag{2.41}$$

where $D_s$ = surface diffusion coefficient, $r$ = radial coordinate, $t$ = time, and $\rho_p$ = pellet density.

This model can be applied to kinetic studies in a batch reactor [53,54], a differential reactor system [19,55], or a fixed-bed adsorber [56,57].

*Pore Diffusion Model*

In the pore diffusion model, it is assumed that the adsorbate diffuses into the pores in the fluid phase and is taken up by adsorption on the walls of the pores. It seems that there are uniformly distributed sinks for adsorption within the pores. The molecule cannot diffuse along the walls of the pores in the adsorbed state but can migrate only by desorbing first. If one assumes (1) a spherical particle, (2) dilute solution, (3) local equilibrium, then the equations which describe the adsorbate mass transfer are as follows:

$$\epsilon_p \frac{\partial C_p}{\partial t} + \rho_p \frac{\partial q}{\partial t} = \frac{\epsilon_p}{r^2} \frac{\partial}{\partial r} \left( r^2 D_p \frac{\partial C_p}{\partial r} \right) \qquad (2.42)$$

$$k_f(C - C_s) = \epsilon_p D_p \frac{\partial C_p}{\partial r} \qquad \text{at } r = R_p \qquad (2.43)$$

$$\frac{\partial C_p}{\partial r} = 0 \qquad \text{at } r = 0 \qquad (2.44)$$

$$C_p = C_{p0}, \qquad q = q_0 \qquad \text{when } t = 0 \qquad (2.45)$$

where $D_p$ = pore diffusion coefficient, $C_{p0}$ = initial solution concentration, and $q_0$ = initial adsorbed amount on solid phase.

Masamune and Smith [58,59] found from their kinetic study that the pore processes control the transfer rate, and the only case where the surface adsorption would affect the adsorption rate is for low site energies. This model was used in the adsorption study for different configurations [60-62].

*Combined Diffusion Model*

Adsorbent particles are heterogeneous systems formed by a porous solid phase and a fluid phase filling the void fraction of the solid. The internal diffusion can be expressed by the two possible simultaneous mechanisms of diffusion, including molecular or Knudsen diffusion, and surface diffusion. If one considers this combined parallel resistance within the adsorbent pellet, a material balance for spherical particle results in the following partial differential equations:

$$\epsilon_p \frac{\partial C_p}{\partial t} + \rho_p \frac{\partial q}{\partial t} = \frac{1}{r^2} \frac{\partial}{\partial r} \left[ r^2 \left( \epsilon_p D_p \frac{\partial C_p}{\partial r} + D_s \rho_p \frac{\partial q}{\partial r} \right) \right] \qquad (2.46)$$

$$k_f(C - C_s) = D_s\rho_p\frac{\partial q}{\partial r} + \epsilon_p D_p\frac{\partial C_p}{\partial r} \qquad \text{at } r = R_p \qquad (2.47)$$

$$\frac{\partial C_p}{\partial r} = 0, \qquad \frac{\partial q}{\partial r} = 0 \qquad \text{at } r = 0 \qquad (2.48)$$

$$C_p = C_{p0}, \qquad q = q_0 \qquad \text{when } t = 0 \qquad (2.49)$$

In the dilute solution range of an adsorption system having a favorable isotherm, Equation (2.46) can be reduced to:

$$\frac{\partial q}{\partial t} = \frac{D_s}{r^2}\frac{\partial}{\partial r}\left(r^2\frac{\partial q}{\partial r}\right) + \frac{\epsilon_p}{\rho_p}\frac{D_p}{r^2}\frac{\partial}{\partial r}\left(r^2\frac{\partial C_p}{\partial r}\right) \qquad (2.50)$$

Application of this model can be found in kinetic studies and fixed-bed adsorption, for gas-solid and liquid-solid systems [63-65].

*REFERENCES*

1. Ruthven, D.M. *Principles of Adsorption and Adsorption Processes.* John Wiley & Sons, New York, NY (1984).
2. Lowell, S. *Introduction of Powder Surface Area.* John Wiley & Sons, New York, NY (1979).
3. Young, D.M., and A.D. Crowell, *Physical Adsorption of Gases.* Butterworths, London, England (1962).
4. Osipow, L.I. "Physical Adsorption on Solids,"Chapter 4 in *Principles and Applications of Water Chemistry: Proceedings of the Fourth Rudolf's Conference.* S.D. Faust and J.V. Hunter, eds. John Wiley & Sons, New York, NY., 75 (1976).
5. London, F. *Trans. Faraday Soc.*, **33**, 8 (1937).
6. Glasstone, S. *Textbook of Physical Chemistry.* Van Nostrand Co., Princeton, NJ (1959).
7. Bokros, J.C. *Chemistry and Physics of Carbon.* Vol.5, P.C. Walker, ed. Marcel Dekker, New York, NY (1969).
8. Mattson, J.S. *Ind. Eng. Chem. Prod. Res. Dev.*, **12**, 312 (1973).
9. Weber, W.J., Jr. "Sorption from Solution by Porous Carbon," Chapter 5 in *Principles and Applications of Water Chemistry.* S.D. Faust and J.V. Hunter, eds. John Wiley & Sons, New York, NY (1967).
10. Langmuir, I.J. *Am. Chem. Soc.*, **37**, 1139 (1915).
11. Smith, A. *Proc. Royal Soc.*, London, **A12**, 424 (1863).
12. Mattson, J.S., and H.B. Mark, Jr., *Activated Carbon.* Marcel Dekker, New York, NY (1971).
13. Kruyt, H.R., and G.S. DeKadt, *Kolloid.*, **2**, 44 (1929).
14. Steenberg, B. *Adsorption and Exchange of Ions on Activated Charcoal.* Almquist & Wiksells, Uppsala, Sweden (1944).
15. Wohleber, D.A., and M. Manes, *J. Phys. Chem.*, **75**, 61 (1971).
16. Michael, R.R., and M. Manes, *J. Phys. Chem.*, **81**, 1651 (1977).
17. Sing, K.S.W. *Pure and Applied Chemistry.*, **54**, 2201 (1981).
18. Andrieu, J., and J.M. Smith, *AIChE. J.*, **26**, 944 (1980).
19. Carlson, N.W., and J.S. Dranoff, *Ind. Eng. Chem. Proc. Des. Dev.*, **24**, 1300 (1985).
20. Ruthven, D.M. *AIChE. J.*, **22**, 753 (1976).
21. Freundlich, H. *Colloids and Capillary Chemistry.* E.P. Dultons and Company, New York, NY (1922).
22. Halsey, T.W. *J. Phys. Chem.*, **36**, 2272 (1947).
23. Langmuir, I.J. *J. Chem. Soc.*, **40**, 1361 (1918).
24. Adamson, A.A. *Physical Chemistry of Surface.* John Wiley & Sons, New York, NY (1982).

25. Itaya, A., N. Kato, J. Yamamoto, and K.I. Okamoto, *J. Chem. Eng. of Japan*, **17**, 389 (1984).

26. Suwanayuen, S., and R.P. Danner, *AIChE. J.*, **26**, 68 (1980).

27. Kondis, E.F., and J.S. Dranoff, *AIChE. Symp.*, **67**, 24 (1971).

28. Myers, A.L., and J.M. Prausnitz, *AIChE. J.*, **11**, 121 (1965).

29. Crittenden, J.C., P. Luft, D.W. Hand, J.L. Oravitz, S.W. Loper, and M. Ari, *Envir. Sci. Tech.*, **19**, 1037 (1985).

30. Yen, C.Y., and P.C. Singer, *ASCE. EE.*, **110**, 976 (1984).

31. Jossens, L., J.M. Prausnitz, W. Fritz, E.U. Schlünder, and A.L. Myers, *Chem. Eng. Sci.*, **33**, 1097 (1978).

32. Crittenden, J.C., B.W.C. Wong, W.E. Thacker, V.L. Snoeyink, and R.L. Hinrichs, *J. WPCF.*, **52**,2780 (1980).

33. Lee, R.G., and T.W. Weber, *Canadian J. Chem. Eng.*, **47**, 54 (1969).

34. Koballa, T.E., and M.P. Dudukovic, *AIChE. Symp.*, **73**, 199 (1977).

35. Wakao, N, and T. Funazkri, *Chem. Eng. Sci.*, **33**, 1375 (1978).

36. Petrovic, L.J., and G. Thodos, *Ind. Eng. Chem. Fundam.*, **7**, 274 (1968).

37. Lightfoot, E.N., R.J. Sanchez-Palma, and D.O. Edwards, *New Chemical Engineering Separation Techniques*. H.M. Schoen, ed., Interscience, New York, NY (1962).

38. Chu, J.C., J. Kalil, and W.A. Wetteroth, *Chem. Eng. Prog.*, **49**, 141 (1978).

39. Ranz, W.E., and W.R. Marshall, *Chem. Eng. Prog.*, **48**, 173 (1952).

40. Wilke, C.R., and O.A. Hougen, *Trans. Am. Inst. Chem. Eng.*, **41**, 445 (1945).

41. Williamson, J.E., K.E. Bazaire, and C.J. Geankoplis, *Ind. Eng. Chem. Fundam.*, **2**, 126 (1963).

42. Hirschfelder, J.O., C.F. Curtiss, and R.B. Bird, *Molecular Theory of Gases and Liquids*. John Wiley & Sons, New York, NY (1954).

43. Wilke, C.R., and P. Chang, *AIChE. J.*, **1**, 264 (1955).

44. Pollard, W.G., and R.D. Present, *Chem. Rev.*, **73**, 762 (1948).

45. Lee, L.K., and D.M. Ruthven, *Ind. Eng. Chem. Fundam.*, **16**, 290 (1977).

46. Wheeler, A. *Catalysis*. Vol.2, P.H. Emmett, ed., Reinhold, New York, NY (1965).

47. Weisz, P.B., and A.B. Schwartz, *J. Catalysis*, **1**, 399 (1962).

48. Johnson, M.F.L., and W.E. Stewart, *J. Catalysis*, **4**, 248 (1965).

49. Froment, G.F., and K.B. Bishoff, *Chemical Reactor Analysis and Design*. John Wiley & Sons, New York, NY (1979).

50. Neretnieks, I. *Chem. Eng. Sci.*, **31**, 465 (1976).

51. Gilliland, E.R., R.F. Baddour, G.P. Perkinson, and K.J. Sladek, *Ind. Eng. Chem. Fundam.*, **13**, 95 (1974).
52. Schneider, P., and J.M. Smith, *AIChE. J.*, **14**, 762 (1968).
53. Roux, A., A.A. Hung, Y.H. Ma, and I. Zwiebe, *AIChE. Symp.*, **63**, 10 (1973).
54. Ma, Y.H., and T.Y. Lee, *AIChE. J.*, **22**, 147 (1976).
55. Sheindorf, C., M. Rebhun, and M. Sheintuch, *Chem. Eng. Sci.*, **38**, 335 (1983).
56. Rosen, J.B. *J. Chem. Phys.*, **20**, 387 (1952).
57. Crittenden, J.C., and W.J. Weber, Jr., *ASCE. EE.*, **104**, 433 (1978).
58. Masamune, S., and J.M. Smith, *AIChE. J.*, **10**, 246 (1964).
59. Masamune, S., and J.M. Smith, *AIChE. J.*, **11**, 41 (1965).
60. DiGiano, F.A., and W.J. Weber, Jr., *J. WPCF.*, **45**, 713 (1973).
61. Svedberg, V.G., *Chem. Eng. Sci.*, **31**, 345 (1976).
62. McKay, G., and M.J. Bino, *Chem. Eng. Res. Des.*, **63**, 168 (1985).
63. Komiyama, H., and J.M. Smith, *AIChE. J.*, **20**, 728 (1974).
64. Komiyama, H., and J.M. Smith, *AIChE. J.*, **20**, 1110 (1974).
65. Cannon, F.S., and P.V. Roberts, *ASCE. EE.*, **108**, 766 (1982).

# CHAPTER III

## THE CONCEPT OF THE DIFFERENTIAL REACTOR

Contributing Author: A.A. Aguwa

### INTRODUCTION

This chapter examines the effective intraparticle diffusion of organic molecules in synthetic resins by using a differential column reactor. The main focus is on the determination of diffusion coefficients at industrial solute concentrations. Specifically, $p$-chlorophenol/XAD-4, $p$-chlorophenol/XAD-2, and phenol/XAD-4 systems were studied. The selection of the two pollutants was based on environmental significance and the adsorbents were chosen for high reversible adsorption. Rate data were collected with a column reactor because the effect of the fluid film resistance on the uptake rate can be eliminated, and the adsorption rate is controlled only by the effective intraparticle diffusion coefficient. A differential reactor offers an advantage over traditional methods (such as completely mixed batch or long columns) in terms of mathematical simplification. While other methods use numerical solutions to estimate effective diffusivity, analytical solutions are easily obtained with a differential column reactor.

### THEORETICAL CONSIDERATIONS

To derive the fundamental equations for the process, it is assumed that the system is isothermal, that there is no resistance from sorption-desorption at the active sites, and that the velocity is high enough to render external mass transfer resistance negligible. The differential reactor assumption requires that the influent and effluent solute concentration be nearly the same. Assuming monodispersed spherical adsorbent particles (closely approximated by most resins) and that equilibrium is represented by a Freundlich isotherm, the following set of equations is obtained from mass balance:

$$\epsilon_p \frac{\partial C_p}{\partial t} + \rho_p \frac{\partial q}{\partial t} = \frac{\epsilon_p}{r^2} \frac{\partial}{\partial r} \left( D_p r^2 \frac{\partial C_p}{\partial r} \right) + \frac{\rho_p}{r^2} \frac{\partial}{\partial r} \left( D_s r^2 \frac{\partial q}{\partial r} \right) \quad (3.1)$$

$$\partial C_p / \partial r = 0 \quad \text{at } r = 0 \quad (3.2)$$

$$\partial q/\partial r = 0 \quad \text{at } r = 0 \tag{3.3}$$

$$C_p = 0 \quad \text{when } t = 0 \tag{3.4}$$

$$q = 0 \quad \text{when } t = 0 \tag{3.5}$$

$$C_p = C_0 \quad \text{at } r = R_p \text{ when } t \geq 0 \tag{3.6}$$

$$q = Q_e \quad \text{at } r = R_p \text{ when } t \geq 0 \tag{3.7}$$

If surface diffusion is assumed to be the main internal transport mechanism, simplification of Equation (3.1) yields:

$$\frac{\partial q}{\partial t} = \frac{1}{r^2}\frac{\partial}{\partial r}\left(D_s r^2 \frac{\partial q}{\partial r}\right) \tag{3.8}$$

If $D_s$ is constant, the system of Equations (3.3), (3.5), (3.7), and (3.8) can be integrated to give [1]:

$$\frac{Q_t}{Q_e} = 1 - \frac{6}{\pi^2}\sum_{m=1}^{\infty}\frac{1}{m^2}\exp\left(-\frac{\pi^2 D_s m^2 t}{R_p^2}\right) \tag{3.9}$$

## EXPERIMENTAL RESULTS AND DISCUSSION

### Materials

The polymeric adsorbents (Amberlite XAD-4 and XAD-2) used in this study have similar chemical structures of styrene-divinylbenzene but different physical properties (Table 1.1). The resin was conditioned by successive washing with distilled water and methanol [2] and stored at 4°C in a sealed container until use. The mean diameter of the resin particles was determined by measuring the diameter of about 300 particles with a microscope. The average particle diameter is 0.388 mm for XAD-4 and 0.3 mm for XAD-2. The weight of the resin was determined after drying at 103°C for 24 hours, and cooling in a dessicator.

The properties of adsorbents, phenol, and p-chlorophenol are listed in Table 3.1. All solutions were prepared by dissolving a known weight of chemical reagent in distilled water. Analysis was performed by ultraviolet (UV) spectrophotometry at wavelengths of 300 and 288 nm for p-chlorophenol and phenol, respectively, in an alkaline medium. The experiments were performed at pH 6.7 ± 0.1 for p-chlorophenol and pH 6.5 ± 0.1 for phenol. At these pH values the molecules are predominantly undissociated.

Table 3.1.   Physical Properties of Adsorbates

| Properties | Phenol | p-Chlorophenol |
|---|---|---|
| Formula | $C_6H_5OH$ | $C_6H_4OHCl$ |
| Molecular weight (g/ml) | 94.11 | 128.56 |
| Solubility (g/l) | 82.0 | 27.1 |
| Heat of solution* (KJ/mol) | 11–13 | 13–16 |
| Ionization constant (pK) | 9.89 | 9.2 |
| Molar volume (cm³/mol) | 96.75 | 112.82 |
| Melting point (⁰C) | 38 | 44 |
| Dipole moment | 1.55–1.73 | 2.22–2.68 |
| Molecular Diffusivity (cm²/sec) | $1.02 \times 10^{-5}$ | $9.32 \times 10^{-6}$ |

\* At $20^0C$ (International Critical Tables, Vol. V).

*Equilibrium Studies*

Equilibrium studies were conducted by adding known quantities of resin and 100-ml aliquots of the solute in distilled water to 125-ml glass bottles, and agitating at $25 \pm 0.5°C$. Equilibrium was attained in about two weeks. The equilibrium concentration of the solute $C_e$ was determined by comparing the amount of UV light adsorbed at the respective wavelengths with the amount of UV light absorbed by a standard solute solution

Figure 3.1 depicts the adsorption isotherms of p-chlorophenol/ XAD-4, phenol/XAD-4, and p-chlorophenol/XAD-2. XAD-4 exhibited higher capacity for p-chlorophenol than XAD-2 did. This was not surprising because both resins have similar chemical structures but XAD-4 has a higher surface area. The capacity for phenol on XAD-4 was lower than p-chlorophenol on XAD-4. The equilibrium data correlated with the Freundlich isotherm, but a statistically significant improvement in the fit was obtained if the equilibrium liquid concentration range was divided into two parts. Correlation coefficients of 0.98 or better were obtained. The Freundlich constants are shown in Table 3.2. The p-chlorophenol/XAD-4 isotherm compares well to that available in the literature [3], where the isotherm data

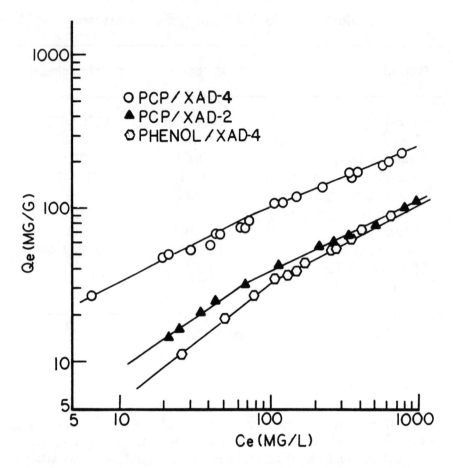

**Figure 3.1.** Isotherms for $p$-Chlorophenol/XAD-4, Phenol/XAD-4, and $p$-Chlorophenol/XAD-2 Systems

was correlated with a four-parameter equilibrium model.

*Kinetic Studies*

The differential reactor used in this study allows measurement of solute accumulation on the solid phase. Because the experimental procedure requires desorption from the solid, extraction efficiency studies were conducted by passing known volumes and concentrations of solute through a differential reactor column (Figure 3.2) and then extracting with sodium hydroxide. 500 ml of 0.1 N NaOH produced quantitative recovery of the solute. Typical extraction efficiencies are shown in Figure 3.3 for the $p$-chlorophenol/XAD-2 system. Similar

Table 3.2.   Constants for the Freundlich Isotherm

| System | Method* | Range (mg/l) | $K$ | $n$ | $R^2$ |
|---|---|---|---|---|---|
| p-Chlorophenol | Batch | Entire | 12.96 | 2.353 | 0.99 |
| /XAD-4 | Batch | $C_e > 88.6$ | 19.02 | 2.774 | 0.98 |
| | Batch | $C_e < 88.6$ | 10.53 | 2.024 | 1.00 |
| | D.R. | Entire | 13.80 | 2.534 | 0.98 |
| Phenol | Batch | Entire | 2.01 | 1.734 | 0.99 |
| /XAD-4 | Batch | $C_e > 100.0$ | 2.84 | 1.920 | 0.99 |
| | Batch | $C_e < 100.0$ | 1.02 | 1.346 | 1.00 |
| | D.R. | Entire | 1.55 | 1.664 | 0.98 |
| p-Chlorophenol | Batch | Entire | 3.23 | 1.970 | 0.99 |
| /XAD-2 | Batch | $C_e > 44.0$ | 3.85 | 2.089 | 0.96 |
| | Batch | $C_e < 44.0$ | 1.61 | 1.409 | 1.00 |
| | D.R. | Entire | 2.25 | 1.846 | 0.98 |

* Batch, from batch reactor; D.R., from differential reactor.

results were obtained for p-chlorophenol/XAD-4 and phenol/XAD-4 systems. In all cases, the loadings during the rate studies were in the range where the extraction efficiencies were 98% or better.

The differential reactor configuration (Figure 3.4) consisted of two 100 liters reservoirs and differential reactor columns. Kinetic experiments involved flowing the solute in water upward through the differential column for a predetermined time interval, removing the column and extracting to determine the amount of solute adsorbed. This procedure was repeated for different columns held at different time periods so that the saturation history of the sorbent could be established.

For the reactor to be regarded as a differential reactor, the concentration gradient in the differential columns must be eliminated by increasing the flowrate through the bed until the influent and effluent concentrations are nearly equal. Because the main objective of this study was to evaluate the effective intraparticle diffusion coefficient, the influent of the fluid film resistance was eliminated by increasing the linear velocity with solute concentration held constant until there

1 GLASS COLUMN    5 EXTRACTION FLASK
2 TEFLON SCREEN    6 COLLECTION FLASK
3 SUPPORT RING     7 STOP COCK
4 PARTICLES BED

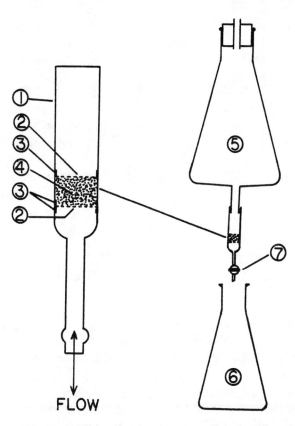

FLOW

**Figure 3.2.**    Differential Bed Reactor and Extraction Apparatus

was a constant uptake versus time profile [4]. Using this procedure, a linear velocity of 13.0 cm/sec was chosen; velocities in excess of this did not materially change the observed rate of adsorption. An example for the $p$-chlorophenol/XAD-4 system is shown in Figure 3.5.

Typical rate data from the differential reactor is shown in Figure 3.6 for the phenol/XAD-4 system. The relationship between the experimental data and model predictions, by using Equation (3.9), is better at low concentrations than at high concentrations. The deviation of the model from experimental points at above 85% saturation

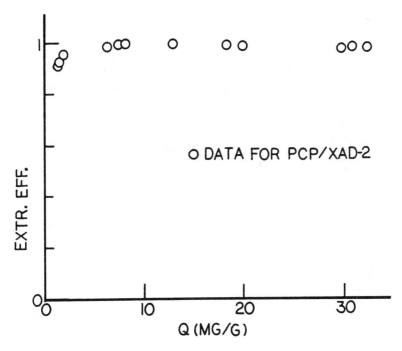

**Figure 3.3.**    Extraction Efficiency versus Amount Adsorbed

occurred in the other systems studied at high solute concentration.

This deviation could be due to concentration and time effects that are not accounted for in the model. Because of the longer contact time and more homogeneous composition in the batch reactor, the adsorption equilibrium capacity so obtained is usually higher than that determined from the fixed bed reactor [5,6]. The slow part of the uptake would practically not be noted in the column sorption experiments, whereas it contributes noticeably in the finite bath experiments which are run for much longer times [7].

Since the differential reactor used in this study is similar to a fixed bed column in terms of the hydrodynamic conditions, it is highly possible that equilibrium capacity determined from the bottle technique is different from the column capacity. The existence of any difference could affect the value of $D_s$ and may explain why the model deviates from experimental data near saturation. To estimate effective diffusivity and column capacity simultaneously, an objective function defined as Equation (3.10) was employed and its minimum value was used as the criteria for defining optimum $D_s$ and $Q_c$.

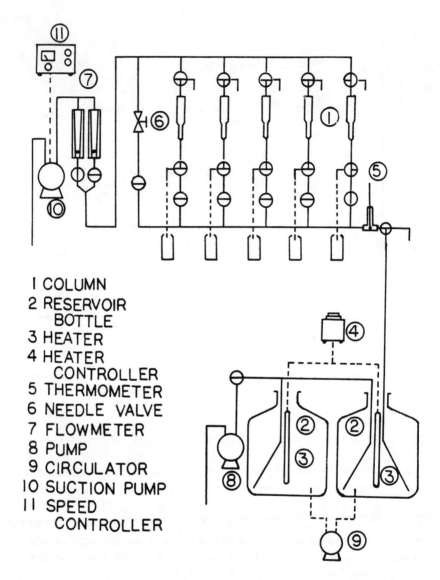

**Figure 3.4.** Schematic Diagram of Experimental Apparatus for Aqueous Solutions

$$\epsilon = \frac{\sum(Q_{t,obs} - Q_{t,cal})^2}{\sum(Q_{t,obs})^2} \tag{3.10}$$

A two-parameters search technique was used in this case. Figure 3.7 shows a typical error map for simultaneous determination of $D_s$ and $Q_c$. This procedure was applied to all kinetic experiments and the

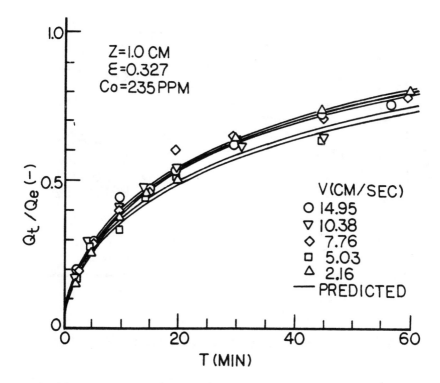

**Figure 3.5.**    Changes in Adsorbed Amount when Flow Rate Varies

$Q_c$'s so obtained were used to analyze the rate data. The equilibrium relationships derived from column capacity are shown in Figure 3.8, and their Freundlich constants are listed in Table 3.2.

To determine the 95% confidence interval of the effective diffusivity, the following procedure was applied for analysis. The average sum of squares of residuals, $\delta^2$, is defined as:

$$\delta^2 = \sum_{i=1}^{n} (\overline{Q}_{exp} - \overline{Q}_{cal})_i^2 / n - 1 \qquad (3.11)$$

where $\delta^2$ = value of the objective function for any given value of $D_s$, $\overline{Q}_{exp}$ = experimental value of $(Q_t/Q_e)$, and $\overline{Q}_{cal}$ = predicted value of $(Q_t/Q_e)$.

This equation based on Equation (3.9) was used where the calculated effective diffusivity is the one for which the objective function is a minimum. The 95% confidence interval was calculated according to the following equation [8]:

$$\delta^2 (95\%, CI) = \delta_m^2 \left[ 1 - \frac{1}{n-1} F(\alpha, n - \alpha, 95\%) \right] \qquad (3.12)$$

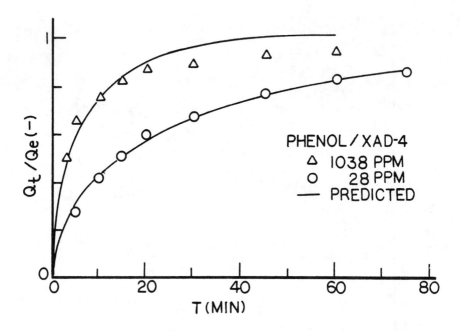

**Figure 3.6.**  Typical Kinetic Curves for Phenol/XAD-4 System

where $\delta^2(95\%, CI)$ = the average sum of squares of the residuals for 95% confidence in $D_s$, $\alpha$ = the number of parameters being determined (one in this case), $\delta_m^2$ = the minimum average sum of squares of the residuals, and $F(\alpha, n - \alpha, 95\%)$ = the distribution function for the 95% confidence interval. In most of the rate experiments, $n = 8$, hence $F(1, 7, 95\%)$ is equal to 5.59 [9]. The rest of the procedure is shown in Figure 3.9. Several values of $D_s$ are assumed and $\delta^2$ calculated and plotted as shown. The $\delta^2(95\%, CI)$ calculated is drawn horizontally to cross the curve at the two points as indicated. This interval defines the 95% confidence interval for $D_s$.

Figure 3.10 shows the effective diffusion coefficient versus external concentration of solute based on the differential reactor data. This figure shows that the effective diffusion coefficient is a function of solute concentration and that the functional relationships depend on the system. For the phenol/XAD-4 system, a linear relationship between effective diffusion coefficient and concentration was found, while for the other systems, curvilinear relationships were observed. Diffusion concentration dependency has been observed by other investigators [10-17] and has been attributed to nonlinearity of the equilibrium isotherm [17-19] and/or changes in the strength of adsorption as evidenced by a change in the differential heat of adsorption [10,15].

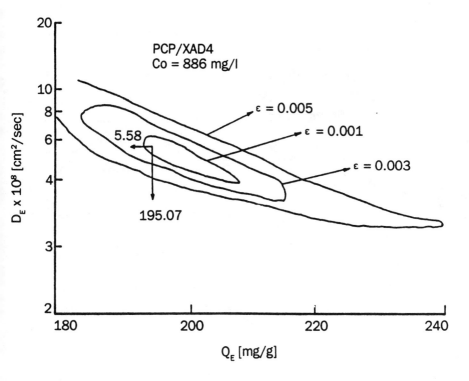

**Figure 3.7.**    Error Map of $D_e$ versus $Q_c$ for a Typical Run

Even though XAD-4 and XAD-2 have similar chemical structures, the difference in physical properties seems to influence the dynamics of the process (Figure 3.10). The kinetic studies reveal that the effective diffusivity is strongly dependent on external solute concentration. This dependency is important for design conditions where the solute concentration is variable or at industrial concentrations ($> 1.0$ mg/l). At environmental concentration ($< 1.0$ mg/l), this dependency on concentration may be less significant. Consequently, the effective diffusivity should be determined using solute concentrations expected during the actual process, if the solute concentration exceeds 1.0 mg/l.

In order to compare the kinetic results between the differential reactor and batch reactor, the differential mixed batch reactor (DMBR) was also used to study adsorption, as shown in Figure 3.11.

The DMBR system consists of (1) a solution reservoir in a water bath where constant temperature is maintained, (2) a differential bed containing the adsorbent, and (3) a stirrer to help maintain uniform sorbate concentration in the reservior. Normal operation consists of inserting the differential column in place and circulating the solution

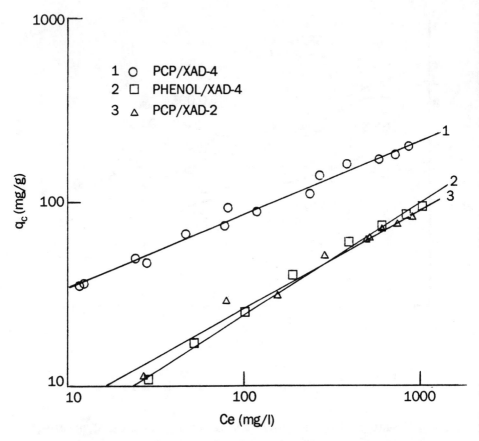

**Figure 3.8.**   Relationship between Column Capacity and Equilibrium Concentration

of known sorbate concentration through the reactor with the effluent returned to the reservoir. The same interstitial velocity used in the differential reactor system was employed in this system. At predetermined time intervals, the reservoir concentration was measured by withdrawing a 2 ml sample and subjecting it to analysis.

The boundary and initial conditions for this DMBR system include Equations (3.3), (3.5), (3.13), and (3.14).

$$\rho_p a_p D_s \frac{\partial q}{\partial r} = -\frac{V_l}{W_s}\frac{\partial C}{\partial t} \qquad \text{at } r = R_p \qquad (3.13)$$

$$C = C_0 \qquad \text{when } t = 0 \qquad (3.14)$$

where $V_l$ = liquid volume, $C$ = liquid concentration, and $W_s$ = weight of adsorbent. Since an analytical solution is not possible for this sys-

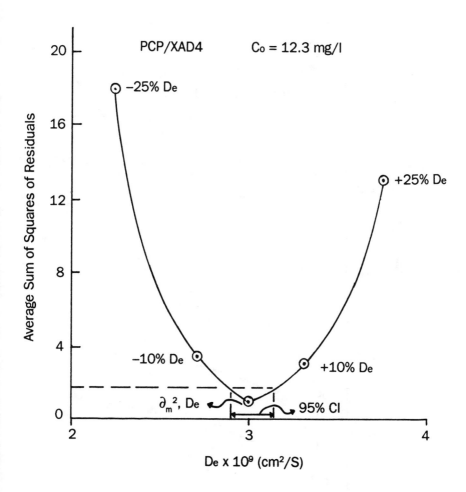

**Figure 3.9.**    Average Sum of Square of Residuals versus Effective
Intraparticle Diffusion Coefficient

tem, those equations were solved numerically by using an orthogonal
collocation method.

Figure 3.12 shows the effective intraparticle diffusion coefficient versus the initial solute concentration for the $p$-chlorophenol/XAD-4 in
the two reactor systems. It can be seen that the DMBR shows a linear
concentration dependency, while the DR systems shows a curvilinear
pattern. This difference can be accounted for by the fact that in the
differential reactor, the concentration remained constant while in the

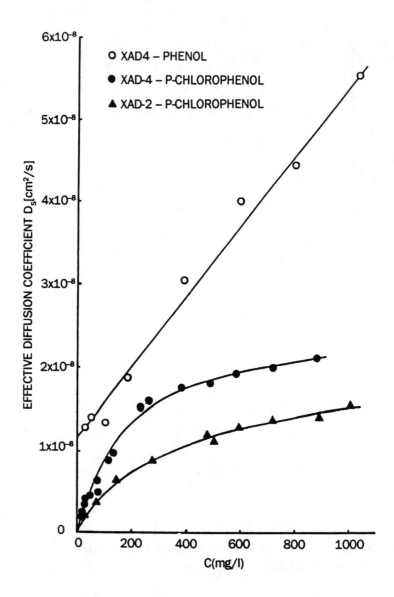

**Figure 3.10.**    Effective Diffusivity versus External Solution
Concentration

DMBR the solution concentration decreased.

Since the isotherm has been implicated by many investigators as
contributing to the concentration dependency of the effective diffusion
coefficient, this was taken into consideration by evaluation, $D_s$ as a
function of $\Delta C/\Delta q$, the reciprocal of the equilibrium isotherm. For

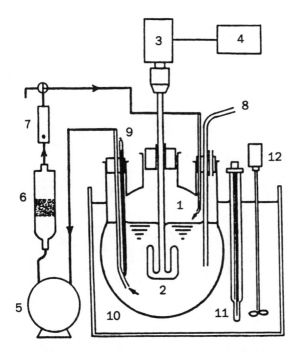

| 1 | REACTOR | 7 | FLOW METER |
|---|---|---|---|
| 2 | STIRRER | 8 | SAMPLING PIPE |
| 3 | MOTOR | 9 | THERMOMETER |
| 4 | SPEED CONTROLLER | 10 | WATER BATH |
| 5 | PUMP | 11 | THERMOCONTROLLER |
| 6 | PARTICLES BED | 12 | STIRRER |

**Figure 3.11.**  Differential Mixed Batch Reactor

the DR system, where surface concentration was maintained constant, $\Delta C / \Delta q$ is simply $C_0 / Q_e$. For The DMBR, the surface concentration varied with time, hence Equation (3.15) was applied.

$$\frac{\Delta C}{\Delta q} = \frac{C_0 - C_e}{Q_e - q_e} \tag{3.15}$$

The data shown in Figure 3.12 is replotted in Figure 3.13 using these concepts. This figure indicates that the two reactor systems give similar results when the operating conditions are taken into consideration.

One of the most important problems derived from this data is that the surface diffusion coefficient $D_s$ is neither uniform throughout the whole particle, nor stable during the entire process of adsorption. This

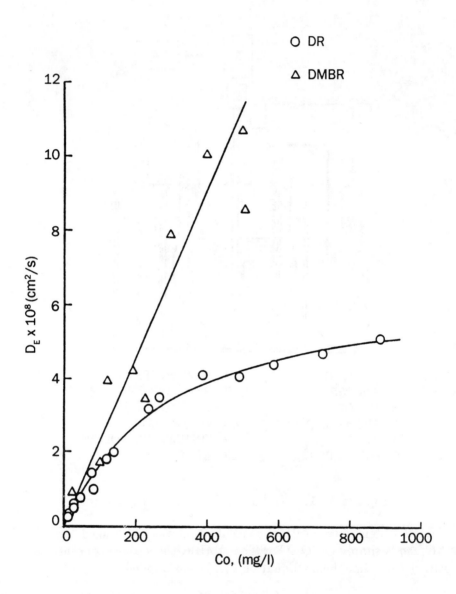

**Figure 3.12.**    Effective Intraparticle Diffusion Coefficient versus Initial Solution Concentration

results from the fact that $D_s$ is strongly dependent on the solid concentration, as shown by Equation (3.16),

$$D_s = D_s^o (aq)^n \tag{3.16}$$

but the adsorbed amount changes continuously with time, as the ad-

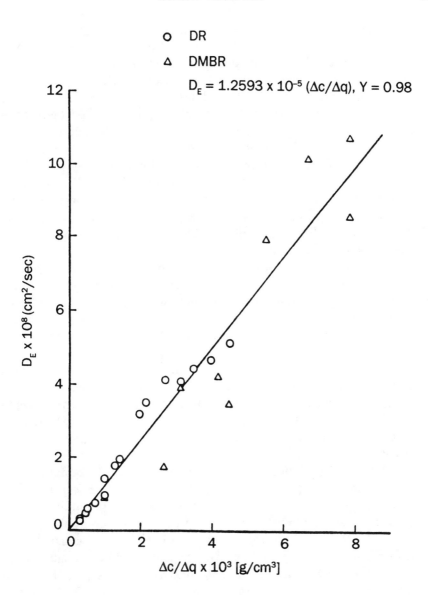

○    DR

△    DMBR

$D_E = 1.2593 \times 10^{-5} \, (\Delta c/\Delta q), \ Y = 0.98$

**Figure 3.13.**    Effective Intraparticle Diffusion Coefficient versus
Reciprocal of Equilibrium Isotherm

sorbate's molecules keep accumulating onto the solid. Furthermore,
the surface of the solid particle is by no means a complete mixed reactor. Thus, even the instantaneous value of solid concentration is
not uniform throughout the entire particle, but varies according to
location. As a result, the value of $D_s$ also varies by location and with

time.

The complex variation of the surface diffusion coefficient poses a serious barrier to the direct application of Equation (3.16) in long column modeling. Even if all the calculations could be handled by advanced numerical methods, the estimation of the solid concentration at a given position, at a given time, would require detailed knowledge of the structure and geometry of the interior of the solid particle. However, even gross characterization of the internal structure of solid adsorbent is difficult at present. The internal tortuosity of carbon, for example, has been reported from 1.0 to 5.0 [20-23]. It is not surprising that computer models developed so far for the prediction of the breakthrough curve of a fixed bed are based on the assumption that this surface diffusion coefficient is constant and independent of the solid phase concentration.

Because the distribution of the sorbate on the solid phase at any time cannot be estimated, the experimental procedure should meet two requirements. First, the experimental loading conditions must be specific and constant for the results to be comparable. Second, these loading conditions should be similar to those existing during the operation of the full-scale unit being modeled.

Those two requirements stem from the fact that different loading conditions produce different solid concentration histories, and thus, different $D_s$ values, even if they result in identical equilibrium conditions. This is illustrated in Figure 3.14, which shows schematically the general distribution patterns of the concentration in a spherical particle exposed to a solution with (a) constant liquid phase concentration, and (b) decreasing liquid concentration (as in a batch configuration). It should be emphasized that the plotted values of concentration in the solid phase and its distribution patterns are purely qualitative and generalized, not displaying the effect of pores of different size.

For both the cases plotted, the value of solid phase concentration is minimized in the central region of the particle. However, although the liquid phase concentration remains constant in case (a), it changes continuously in case (b), starting from a high initial value and dropping gradually to the value at final equilibrium. These changes have to be closely followed by the superficial local solid phase concentration, while the response of the more internal parts of the particle is slower and less extensive.

The result of the above differences is that at an early time $t_1$ the gradient of the solid phase concentration is higher in case (b) than in (a). The same is true for the instantaneous average over the particle solid phase concentration. At a later time $t_2$, the distribution of the

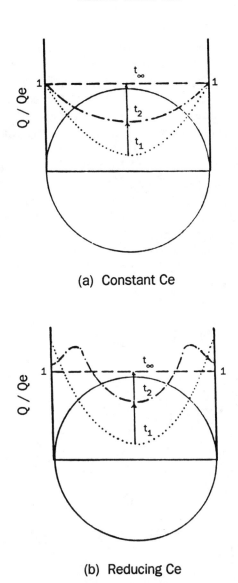

(a) Constant Ce

(b) Reducing Ce

**Figure 3.14.** Distribution of Concentration in Solid Phase

solid phase concentration in case (a) is quite different from the one in case (b). After long enough time $t$, the final equilibrium conditions are identical. However, they have been achieved by two completely different histories of solid phase concentration. This means that the particular instantaneous local values of $D_s(r)$ were also different between the two cases. Consequently, the resulting average $D_s$ values may also be different.

Despite their discrepancy with theory, those models have been successful in predicting the breakthrough curve of an operating column provided a suitable $D_s$ value is used. This means that the whole process of surface diffusion can be described, at the level of application, in terms of an average value of the surface diffusion coefficient. This average value $D_s$ represents the whole adsorbent particle, and the entire duration of the adsorption process.

For fixed-bed design or modeling, $D_s$ must be determined from rate studies since there is no general correlation. The lack of correlation is believed to be due to a lack of knowledge and characterization of the sorbent and quantification of all interactions. An attempt was made to correlate all the data obtained in this study for different solute and sorbent, as shown in Figure 3.15. Plotted is dimensionless diffusion parameter $\overline{D}$ versus dimensionless equilibrium parameter $\overline{E}$, as shown by the following equations:

$$\overline{D} = \overline{D}_0 \overline{E}^S \tag{3.17}$$

$$\overline{D} = \frac{D_s V_A}{D_m A_s M r_p} \tag{3.18}$$

$$\overline{E} = \frac{\Delta C}{\Delta q \rho_s} \tag{3.19}$$

For this particular application, $\overline{D}_0$ is equal to 0.6847; $S$ is equal to 1.046 and statistically not different from unity. Therefore, Equation (3.17) may be rewritten as Equation (3.20).

$$\overline{D} = \overline{D}_0 \overline{E} \tag{3.20}$$

Equation (3.20) simply says that the diffusion parameter is directly proportional to the equilibrium parameter. Since these parameters are dimensionless, the correlation may be used for the estimation of $D_s$ in sorbents with similar characteristics, or Equation (3.20) can be used as a starting point to establish a comprehensive correlation equation applicable to all systems. To achieve this goal, the sorbent must be studied in detail and be well characterized.

## CONCLUSIONS

Rate studies conducted at various solute concentrations and with two reactor configurations show strong dependency of the effective intraparticle diffusion coefficient on solute concentration. Since the

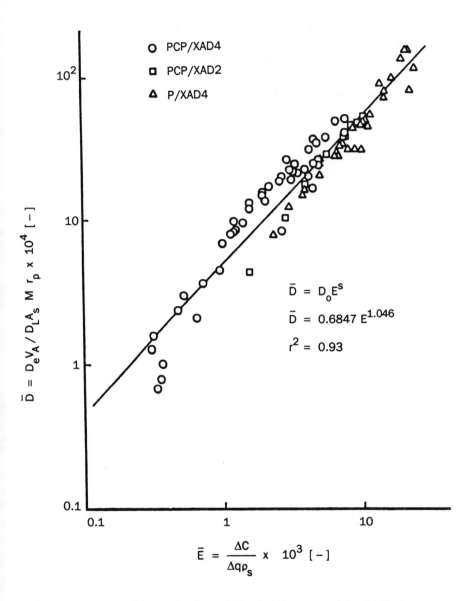

**Figure 3.15.** Dimensionless Effective Intraparticle Diffusion Coefficient versus Dimensionless Equilibrium Parameter

state-of-the-art in fixed bed adsorber mathematical modeling or design uses constant diffusion coefficient, it is highly recommended that the rate studies be conducted at the anticipated concentration range. It is also shown that the use of a differential reactor simplifies the complex mathematics associated with the measurement and modeling of the effective intraparticle diffusivity, therefore, its use is encouraged.

Finally, it is shown that the differential reactor results are similar to the differential mixed batch reactor or the traditional completely mixed batch reactor if and only if the operating conditions due to different liquid concentrations in the DMBR are taken into consideration.

*REFERENCES*

1. Crank, J. *The Mathematics of Diffusion.* Oxford University Press, London, England (1956).
2. Rohm and Haas Co. *Technical Bulletin – Amberlite XAD-4 and XAD-2.* Rohm & Haas Technical Bulletin, Philadelphia, PA (1978).
3. Weber, W.J., Jr., and B.M. Van Vliet, *J. WPCF.* **53**, 1585 (1981).
4. Noll, K.E. *Direct Differential Reactor Studies on Adsorption from Industrial Strength Liquid and Gaseous Solutions.* Presentation at Engineering Foundation Conferences (Fundamentals of Adsorption), Schloss Elmau, Bavaria, West Germany (1983).
5. Johansson, R., and I. Neretnieks, *Chem. Eng. Sci.*, **35**, 979 (1980).
6. McKay, G., and M.J. Bino, *Chem. Eng. Res. Des.*, **63**, 168 (1985).
7. Hashimoto, K., K. Miura, and M. Tsukano, *J. Chem. Eng. Japan,* **10**, 27 (1977).
8. Draper, N.R., and H. Smith, *Applied Regression Analysis.* John Wiley and Sons, New York, NY (1981).
9. Dunn, O.J., and V.A. Clark, *Applied Statistics: Analysis of Variance and Regression.* John Wiley and Sons, New York, NY (1974).
10. Gilliland, E.R., R.F. Baddour, G.P. Perkinson, and K.J. Sladek, *Ind. Eng. Chem. Fundam.*, **13**, 95 and 100 (1974).
11. Suzuki, M., and T. Fujii, *AIChE. J.*, **28**, 380 (1982).
12. Kruckels, W.W. *Chem. Eng. Sci.*, **28**, 1565 (1973).
13. Higashi, K., H. Ito, and J. Oishi, *J. Atomic Energy Soc. Japan*, **5**, 846 (1963).
14. Okazaki, M. *AIChE. J.*, **27**, 262 (1981).
15. Neretnieks, I. *Chem. Eng. Sci.*, **31**, 1029 (1976).
16. Frost, A.C. *AIChE. J.*, **27**, 813 (1981).
17. Fritz, W., W. Merk, and E.U. Schlünder, *Chem. Eng. Sci.*, **36**, 721 (1981).
18. Ruthven, D.M., and K.F. Loughlin, *Trans. Fara. Soc.*, **67**, 1661 (1971).
19. Garg, D.R. and D.M. Ruthven, *Chem. Eng. Sci.*, **27**, 417 (1972).
20. Satterfield, C.N. *Mass Transfer in Heterogeneous Catalysis.* M.I.T. Press, Cambridge, Mass. (1970).
21. Costa, E., G. Calleja, and F. Domingo, *AIChE. J.*, **31**, 982 (1985).
22. Andrieu, J., and J.M. Smith, *AIChE. J.*, **26**, 944 (1980).
23. Suzuki, M., and K. Kawazoe, *J. Chem. Eng. of Japan*, **7**, 346 (1974).

# CHAPTER IV

## GRAVIMETRIC DIFFERENTIAL REACTOR FOR GAS ADSORPTION STUDIES

Contributing Author: J.N. Sarlis

### INTRODUCTION

Usually, for laboratory studies of equilibrium and rate behaviors, experiments are conducted through gravimetric [1-3], columnar [4,5], or diffusion cell [6] reactors. Although the experimental procedures are relatively simple, complex mathematical expressions must be solved numerically to obtain the rate parameters.

This chapter investigates the use of a gravimetric differential reactor, which allows the simultaneous determination of rate and equilibrium information [7] and simplifies the mathematical solutions. Adsorption phenomena of volatile organic compounds (industrial solvents), are compared by examining their sorptive behavior on three selected adsorbents. The six organic vapors were chosen for their environmental significance and their wide range of physicochemical characteristics [8].

### EXPERIMENTAL RESULTS AND DISCUSSION

#### Materials

The three adsorbents used are Kureha beaded activated carbon BAC, molecular sieve 13X, and Amberlite copolymer resin XAD-4; their properties are listed in Table 1.1. Those adsorbents have been widely used and have high selectivity in capturing organic vapors. The particle size distribution based on 300 randomly selected particles is 0.0674 cm for BAC, 0.1826 cm for MS-13X, and 0.0462 cm for XAD-4.

The organic vapors selected were toluene ($C_7H_8$), $p$-xylene ($p$-$C_8H_{10}$), carbon tetrachloride ($CCl_4$), benzene ($C_6H_6$), tetrachloroethylene ($C_2Cl_4$) and trichloroethylene ($C_2HCl_3$). The gaseous vapors were generated from their respective liquid forms. The pure liquids were 99% pure. Xylene was used in its para-xylene form because of its higher purity. The principal properties of these gases are listed in Table 4.1. The range of concentrations investigated was limited by

Table 4.1.  Physical Properties of Adsorbates [9]

| Property | $C_6H_6$ | $C_7H_8$ | $p\text{-}C_8H_{10}$ | $CCl_4$ | $C_2HCl_3$ | $C_2Cl_4$ |
|---|---|---|---|---|---|---|
| Molecular weight | 78.01 | 92.12 | 106.16 | 153.82 | 131.39 | 165.83 |
| Liquid density $25^0C$ (g/ml) | 0.88 | 0.867 | 0.861 | 1.584 | 1.462 | 1.620 |
| Boiling point ($^0C$) | 80.30 | 110.6 | 138.35 | 76.70 | 87.40 | 121.30 |
| Explosion limit (vol %) | 1.4–7.1 | 1.27–6.75 | 1.0–6.0 | n.f. | n.f. | n.f. |
| Vapor pressure $23^0C$ (atm) | 0.1133 | 0.0335 | 0.0102 | 0.1365 | 0.0883 | 0.0216 |
| Heat of condensation* (Kcal/mole) | 7.90 | 8.75 | 9.62 | 7.63 | 7.99 (8.04)[†] | 9.03 |

n.f.: nonflammable
*: Calculated by the Clausius-Clapeyron Equation
†: Perry's Chemical Engineers Handbook [10]

the saturation vapor pressures of the organic vapors.

*Procedure*

The experimental apparatus is depicted in Figure 4.1 and consists of four main parts: (1) vaporizer for the generation of organic vapor, (2) temperature controllers for both the vaporizer and the adsorption column, (3) the adsorption column where a teflon basket containing the sorbent is hung from a quartz spring, and (4) a cathetometer for measuring the spring expansion caused by the organic vapor adsorption onto the sorbent.

The experimental procedure consisted of placing a monolayer of adsorbent particles in the teflon basket and attaching it to the quartz spring. The organic vapor was then generated by pumping the pure organic liquid at a desired flow rate into the vaporizer. Once vaporized, the organic vapor was carried to the adsorption column housing the basket by a nitrogen carrier gas. The average adsorbent weight in the basket was approximately 0.04 g and the gas mixture linear velocity was constant at 35.4 cm/sec. The column temperature was maintained at 23±1°C throughout all the studies.

As the mixture of the organic vapor and nitrogen passed through the adsorption column, adsorption of the organic vapor onto the adsorbent resulted in the expansion of the quartz spring. The extent of expansion was determined by a cathetometer. The extension coefficient of the spring was found to be 10.3327 mg/cm. Figure 4.2 shows the calibration curve for the quartz spring. The precision of the reading of the expansion of the spring with the cathetometer was 0.01 cm.

Because of the adsorption of the organic vapor, the spring expands until equilibrium is reached. Thus, both the rate of adsorption and equilibrium behavior can be studied simultaneously. At any desired time during the adsorption process the flow can be momentarily diverted to measure the expansion and hence the uptake. This process was repeated for various time intervals to produce the saturation history of the organic vapor on the adsorbent. Equilibrium was assumed when the spring no longer expanded.

Due to the monolayer procedure and because of the flow rate used, this type of reactor can be termed a differential reactor, i.e., the concentration before and after the resin are approximately the same [11,12]. Additionally, the high flow rate justifies the negligence of the fluid film resistance, thus only intraparticle resistance was assumed for the interpretation of the rate data.

*Equilibrium Isotherm*

Figures 4.3–4.5 depict the adsorption isotherms for the various experimental systems examined in this study. All equilibrium relationships obtained in this work were nonlinear, favorable isotherms which could be correlated by the Freundlich model.

$$Q_e = K P^{1/n} \tag{4.1}$$

Freundlich parameters are shown in Table 4.2.

All systems consistently show that activated carbon has a higher terminal equilibrium capacity than XAD-4 and molecular sieves. The

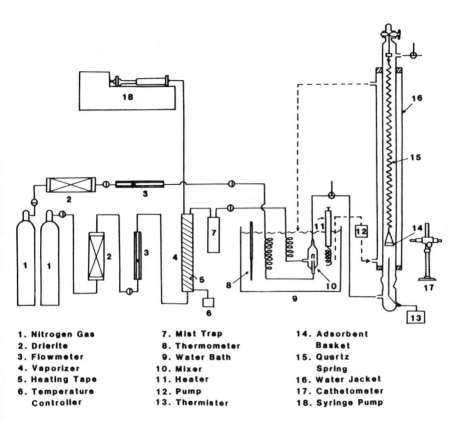

| 1. Nitrogen Gas | 7. Mist Trap | 14. Adsorbent |
| 2. Drierite | 8. Thermometer | Basket |
| 3. Flowmeter | 9. Water Bath | 15. Quartz |
| 4. Vaporizer | 10. Mixer | Spring |
| 5. Heating Tape | 11. Heater | 16. Water Jacket |
| 6. Temperature | 12. Pump | 17. Cathetometer |
| Controller | 13. Thermister | 18. Syringe Pump |

**Figure 4.1.**   Experimental Apparatus for Organic Vapor
Adsorption System

slope of the isotherms for XAD-4, are much higher than those for activated carbon systems. As a result, the isotherms will cross at some point in the range of industrial concentrations (greater than 25 percent). Furthermore, when considering the high reversibility of the polymeric resin systems (i.e., recovery of solvents) as well as their fast kinetics, the value of resins for adsorption in the solvent industry can be understood.

Full range equilibrium values for industrial solvents and volatile organic compounds are scarce in the literature. The data produced by Sydor and Pietrzyk [3] with benzene on XAD-4 portray an isotherm statistically similar to the one obtained in this study as shown in Figure 4.4. The authors employed a continuous flow sampling system

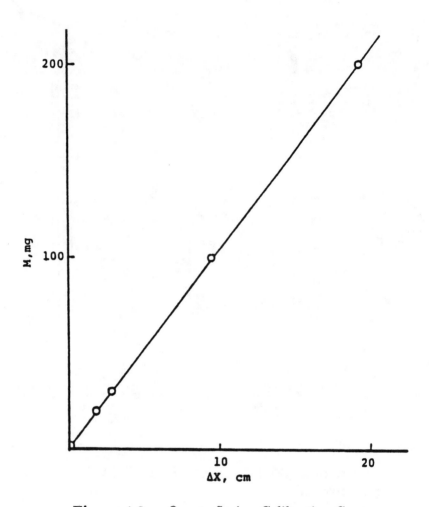

**Figure 4.2.** Quartz Spring Calibration Curve

consisting of a small sampling tube. The different sampling methods are considered responsible for the deviation in the region of 30 ppm. The U.S. Environmental Protection Agency [13] reports several carbon tetrachloride isotherms on oil pitch activated carbon at various temperatures. Figure 4.4 depicts the characteristics of the closest representation to the system found in this study. The isotherm was derived at 25°C (as opposed to 23° in this work) and in a long column. The small difference in equilibrium capacity might be attributed to non-isothermal heat effects and wall-proximity flow patterns, as well as differences in the two carbons used.

Urano et al. [14] report equilibrium values of benzene, toluene, o-

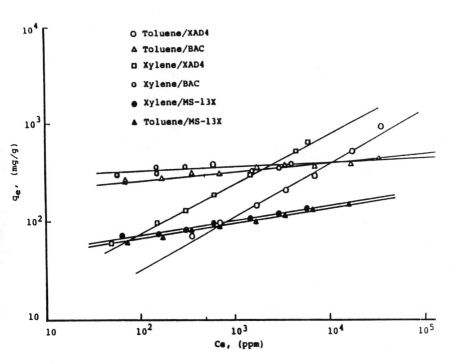

**Figure 4.3.** Equilibrium Isotherms of $C_7H_8$ and $p$-$C_8H_{10}$ on BAC and MS-13X

xylene, carbon tetrachloride and trichloroethylene on two different kinds of coconut shell activated carbon with about 17 percent more surface area than the activated carbon used in this study. The reported equilibrium values agree to within ± 15 percent or less of the values found in this investigation. Once again, the differences can be attributed to the distinct characteristics of the adsorbents and the difference in the temperatures against which the isotherms were drawn. Figure 4.3 depicts the toluene and $p$-xylene isotherms on molecular sieves 13X.

*Kinetic Rate Studies*

If the system is assumed to be isothermal, there is no resistance due to adsorption at the active sites, there is negligible external mass transfer resistance resulting from high velocity, and local equilibrium

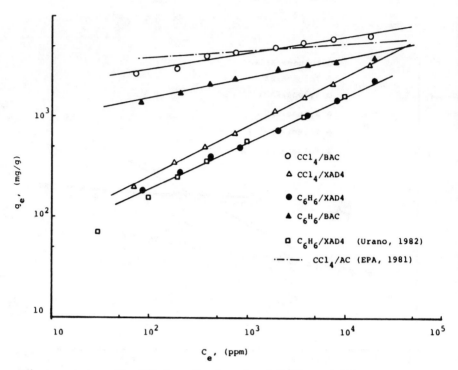

**Figure 4.4.** Equilibrium Isotherms of $C_6H_6$ and $CCl_4$ on XAD-4 and BAC

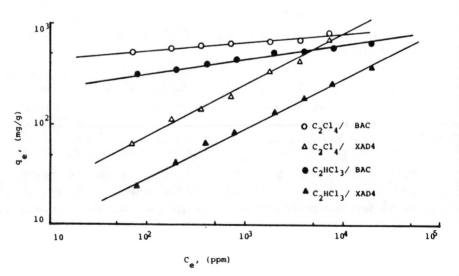

**Figure 4.5.** Equilibrium Isotherms of $C_2Cl_4$ and $C_2HCl_3$ on XAD-4 and BAC

Table 4.2.    Freundlich Parameters for Various Isotherms

| System | $K$ | $n$ | $R^2$ |
|---|---|---|---|
| Toluene/XAD-4 | 3.37 | 1.94 | 0.99 |
| Toluene/BAC | 178.41 | 11.74 | 0.96 |
| Toluene/MS-13X | 29.16 | 5.88 | 0.99 |
| p-Xylene/XAD-4 | 6.70 | 1.95 | 0.99 |
| p-Xylene/BAC | 258.72 | 20.20 | 0.72 |
| p-Xylene/MS-13X | 34.39 | 6.25 | 0.90 |
| Benzene/XAD-4 | 2.06 | 2.13 | 0.98 |
| Benzene/BAC | 62.01 | 5.26 | 0.93 |
| Carbon tetrachloride/XAD-4 | 2.51 | 2.00 | 0.99 |
| Carbon tetrachloride/BAC | 141.57 | 6.25 | 0.95 |
| Trichloroethylene/XAD-4 | 2.88 | 2.00 | 0.99 |
| Trichloroethylene/BAC | 158.42 | 6.67 | 0.94 |
| Tetrachloroethylene/XAD-4 | 7.63 | 2.00 | 0.99 |
| Tetrachloroethylene/BAC | 353.25 | 11.11 | 0.94 |

exists at the interphase Equation (2.50) is applicable for combined surface and pore diffusion [15]. Equation (2.50) can be simplified as follows:

$$\frac{\partial q}{\partial t} = \frac{D_e}{r^2}\frac{\partial}{\partial r}\left(r^2\frac{\partial q}{\partial r}\right) \tag{4.2}$$

$$D_e = D_s + \frac{D_p}{\rho_p}\frac{\partial C_p}{\partial q} \tag{4.3}$$

where $D_e$ = effective intraparticle diffusion coefficient. Note that $\partial C_p/\partial q$ is the reciprocal of the partial derivative of the isotherms. For a linear isotherm $\partial C_p/\partial q$ is constant but for a nonlinear isotherm there is no simple relationship, hence, $\partial C_p/\partial q$ may be estimated as $C_e/Q_e$ [16]. Therefore, the contribution of surface diffusion and pore diffusion in the intraparticle diffusion can be estimated.

For the differential reactor system, the external concentration is nearly constant during the process, so that an analytic solution can be derived as [17]:

$$\frac{Q_t}{Q_e} = 1 - \frac{6}{\pi^2} \sum_{m=1}^{\infty} \frac{1}{m^2} \exp\left(-\frac{\pi^2 D_e m^2 t}{R_p^2}\right) \qquad (4.4)$$

Figure 4.6 depicts a typical example of adsorption rate data obtained during the process. Plotted in the figure is the normalized amount adsorbed per mass of sorbent versus time.

Equation (4.4) was used in the analysis of rate data and subsequent determination of the intraparticle diffusion coefficient. A curve-fitting procedure was used and the aim was to minimize the deviations between the observed $Q_t$ values and those calculated by the model. Consequently, an objective function, $\epsilon$, was defined as:

$$\epsilon = \frac{\int (Q_{t,obs} - Q_{t,cal})^2 \, dt}{\int Q_{t,obs}^2 \, dt} \qquad (4.5)$$

where $Q_{t,obs}$ = the amount of adsorbate adsorbed and $Q_{t,cal}$ = the amount of adsorbate calculated from the model (see Equation (3.10)). The minimum of the objective function defines the value of the effective intraparticle diffusion coefficient. This method provides an optimum equilibrium uptake $Q_e$ and an optimum effective diffusivity $D_e$. The best-fit curve using the optimum intraparticle diffusion coefficient is also shown in Figure 4.6. The best-fit uptake curve agrees reasonably well with the observed values.

Figures 4.7–4.9 show plots of effective diffusivity $D_e$ versus the vapor concentration $C_e$ when $D_e$ was assumed to be constant throughout each kinetic experiment. The main feature of these types of plots is that the effective intraparticle diffusivities are concentration dependent. Consequently, no universal value for $D_e$ may be used for the design of fixed beds since $D_e$ depends on the adsorbate concentration. Concentration dependency of intraparticle diffusion coefficients has been observed by many investigators and has been attributed to the nonlinearity of the equilibrium isotherm [2,18] and/or changes in the strength of adsorption as evidenced by a change in the differential heat of sorption [19,20].

Garg and Ruthven [18] produced a family of curves comparing the analytical solutions to Equation (3.8), assuming constant diffusivity $D_e$, with exact numerical solutions of the same equation without the assumption of constant effective diffusivity for various values of the degree of relative sorbate saturation $\lambda$. The relative degree of saturation was varied from 0 to 100 percent by volume. Their conclusion was

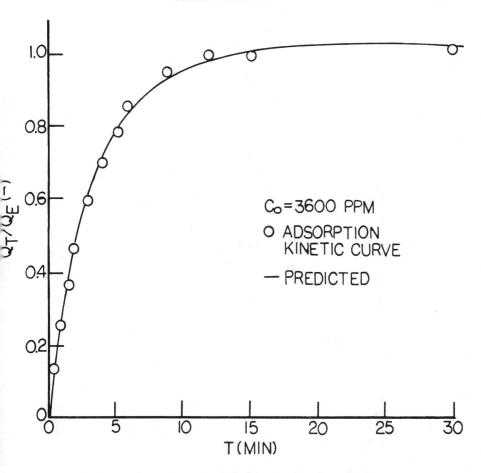

**Figure 4.6.** Experimental and Predicted Uptake Curve of
$C_2Cl_4/BAC$

that for $0 < \lambda < 0.1$, there is no practical or mathematical difference
in the final solution of Equation (3.8). For values of $0.1 < \lambda < 0.3$,
the constant diffusivity model is a good approximation to the exact
numerical solution (less than 2 percent difference). Throughout this
study, all sorbate concentrations were smaller than 30 percent by volume. Hence, it is believed that the intraparticle diffusion coefficients
obtained in this work are very close to those one can obtain with very
laborious numerical solutions of the nonconstant diffusivity Fickian
model.

*CONCLUSIONS*

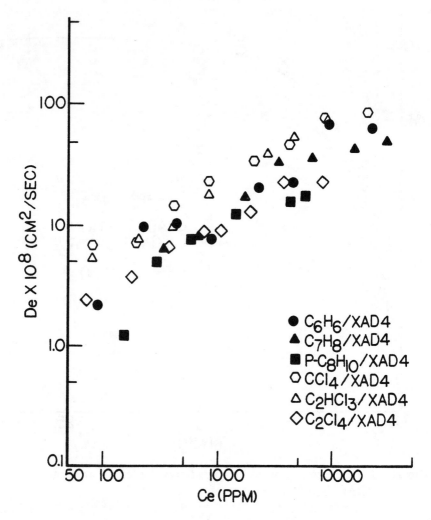

**Figure 4.7.**    Effective Diffusivity versus Concentration (XAD-4)

The objective of this study was to evaluate the sorptive behavior of six hazardous organic solvents on activated carbon, XAD-4 resin, and molecular sieve 13X. The adsorbates were chosen from among the many hazardous organic solvents to provide a wide range of gaseous physicochemical properties.

The experimental procedure allowed simultaneous study of both equilibrium and kinetic conditions. Different results were obtained for each of the six solutes (organic vapors) on XAD-4 resin and activated carbon, and this is believed to be due to the different physical/chemical properties of the solutes and the sorbents. The equilib-

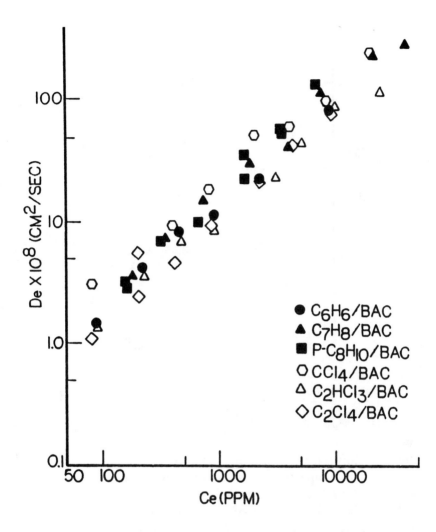

**Figure 4.8.** Effective Diffusivity versus Concentration (BAC)

rium studies showed higher capacity of the adsorbates on activated carbon at low concentrations, but the consistently larger slopes of the isotherms for XAD-4 indicate that at higher industrial level concentrations the resin had a higher capacity.

The effective diffusivities were found to be concentration-dependent. This characteristic was attributed to the nonlinearity of the sorption isotherm which suggests that no universal value of $D_e$ can be used for the design of sorption equipment. The rate studies also showed that the surface diffusivity $D_s$ is concentration-dependent. These results

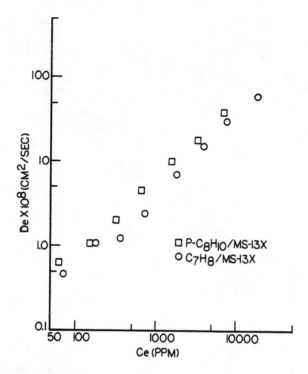

**Figure 4.9.** Effective Diffusivity versus Concentration (MS-13X)

are similar to those developed for liquid-solid adsorption systems in Chapter III.

Effective diffusivities estimated were of the order $10^{-7}$ to $10^{-8}$ cm$^2$/sec which compares well with the range of diffusivities for organic vapors found in the literature. The difference in the adsorption behavior of the three different adsorbents is due in part to the difference in the pore structure and material characteristics.

*REFERENCES*

1. Ruckerstein, E. *Chem. Eng. Sci.*, **26**, 1305 (1971).
2. Ruthven, D.M., and K.F. Loughlin, *Trans. Fara. Soc.*, **67**, 1661 (1971).
3. Sydor, R., and D.J. Pietrzyk, *Anal. Chem.*, **50**, 1842 (1978).
4. Chihara, K. *AIChE. J.*, **24**, 237 (1978).
5. Schneider, P., and J.M. Smith, *AIChE. J.*, **14**, 886 (1968).
6. Frost, A.C. *AIChE. J.*, **27**, 813 (1981).
7. Noll, K.E., A.A. Aguwa, Y.P. Fang, and P.T. Boulanger, *ASCE. EE.*, **111**, 487 (1985).
8. Sarlis, J.N. *Sorptive Behavior of Hazardous Organic Solvents by Direct Differential Reactor Measurements.* M.S. Thesis, Illinois Institute of Technology, Chicago, IL (1985).
9. Reid, R.C., T.M. Prausnitz, and T.K. Sherwood, *The Properties of Gases and Liquid.* McGraw-Hill, New York, NY (1977).
10. Perry, J.H. *Chemical Engineering Handbook.* 5th ed., McGraw-Hill, New York, NY (1983).
11. Boulanger, P.T. *Adsorption of Organic Vapors on Synthetic Resin.* M.S. Thesis, Illinois Institute of Technology, Chicago, IL (1984).
12. Fang, Y.P. *Adsorption and Desorption Study of Toluene Vapor on XAD-4 Resin Using Single Particle Layer Method.* M.S. Thesis, Illinois Institute of Technology, Chicago, IL (1984).
13. U.S. EPA. *Control of Gaseous Emissions Manual.* Research Triangle Park, NC (1981).
14. Urano, K., S. Omori, and E. Yamamoto, *Env. Sci. Tech.*, **16**, 10 (1982).
15. Kobala, T.E., and M.P. Dudukovic, *AIChE. Symp.* **73**, 199 (1977).
16. Neretnieks, I. *Chem. Eng. Sci.*, **31**, 1029 (1976).
17. Crank, J. *The Mathematics of Diffusion.* Oxford University Press, London, England (1956).
18. Garg, D.R., and D.M. Ruthven, *Chem. Eng. Sci.*, **27**, 417 (1972).
19. Gilliland, E.R., R.F. Baddour, G.P. Perkinson, and K.J. Sladek, *Ind. Eng. Chem. Fundam.*, **13**, 95 (1974).
20. Sircar, S. *Carbon*, **4**, 285 (1980).

# CHAPTER V

---

## GRAVIMETRIC DIFFERENTIAL REACTOR
## FOR WATER ADSORPTION STUDIES

---

### INTRODUCTION

A gravimetric method using a quartz spring was applied to study the adsorption equilibrium and kinetics in liquid/solid systems. This experimental technique has been used many times in the gas/solid system for equilibrium studies [1-5] and for kinetic studies [6] (see Chapter IV). However, due to the problem of buoyancy, the interpretation of the experimental data are more complicated in the water/solid system than in the gas/solid system. In this study, the buoyancy factor was found to be constant for each adsorption system and was determined experimentally.

### EXPERIMENTAL RESULTS AND DISCUSSIONS

*Experimental Procedure*

The adsorbents Filtrasorb 400 and Amberlite XAD-4 and the adsorbates phenol and $p$-chlorophenol are used in the experiments. The adsorbent particle size, after sieving and microscopic analysis, was found to be 0.45 mm for XAD-4 and 0.78 mm for F400. The properties of the adsorbents are listed in Table 1.1 and that of the adsorbates are shown in Table 3.1. Figure 5.1 depicts the structure of adsorbents XAD-4 and F400.

The apparatus consists of a Quartz Spring Balance in a air-filled column and a differential reactor in a liquid-filled column. The extension coefficient of the spring was found to be 10.0 mg/cm over the range of operation. The differential reactor, a Teflon basket of 1.1 cm OD, 0.9 cm ID, and 0.5 cm length, containing about 0.2 g adsorbent, was immersed in the liquid phase. The reactor was suspended from the spring by a glass wire which is thin enough ($d < 0.05$ mm) to ignore the buoyance effect in water. The parts in contact with the organic solution were made of glass or Teflon, as shown in Figure 5.2. Before each run the system was flushed with distilled water to re-

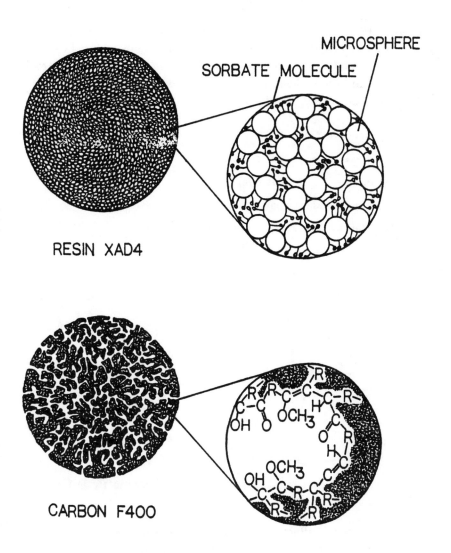

**Figure 5.1.** Structure of Adsorbents XAD-4 and F400

move impurites and any fine air bubbles from the reactor. During the experiment, an upward stream of solution was pumped through the column, so that the basket drifted upward to fit the narrow top of the column and contacted the total flow. Whenever a measurement was made, the flow was stopped and the suspended basket was allowed to balance in the stagnant liquid phase. The expansion of the spring, that resulted from the adsorbed amount in the basket, was measured by using a cathetometer, which can be read to 0.01 cm. The effluent from the reactor was recycled to a reservoir in which the volume of the organic solution was about 150 liters. Due to the large solution volume relative to the small adsorbent amount, the concentration change in the liquid phase by adsorption was less than 0.1%, so that the concentration was assumed to be constant.

① Solution Tank    ⑦ Temp. Controller

② Water Bath    ⑧ Thermometer

③ Pump    ⑨ Basket

④ Flow Meter    ⑩ Cathetometer

⑤ Flow Controller    ⑪ Quartz Spring

⑥ Sampling Point    ⑫ Discard

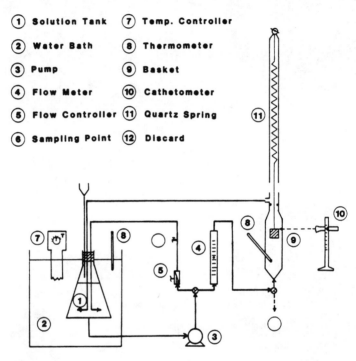

**Figure 5.2.** Experimental Apparatus of Gravimetric Differential Reactor

To eliminate the effect of the liquid film resistance, a linear velocity of 25 cm/sec in the reactor was selected for the experiments. This velocity was determined by increasing the flow rate until the uptake versus time profile remained constant [7,8]. Due to this high flow rate, equilibrium can generally be achieved in 24 hours.

*Buoyancy Determination*

Experimental Method. The adsorbed amount in the resin XAD4 can be determined by extraction using 0.1 N NaOH solution [7]. Results showed that the measured values by gravimetry were systematically lower than by extraction, as shown in Figures 5.3 and 5.4. However, the ratio of the two measurements was found to be constant for all the concentrations measured. This ratio was 1.9 with deviation 0.2 for phenol/XAD-4, and 1.82 with deviation 0.17 for p-chlorophenol/XAD-4. Due to this relation, the bouyancy factor $A$ was introduced into the gravimetric system:

$$W = ARE \qquad (5.1)$$

where $E$ = extension of the spring; $R$ = constant of the spring (force/length) and $W$ = weight of adsorbed molecules.

The force balance over the adsorbent is as follows:

$$ER = W - B \qquad (5.2)$$

where $B$ = buoyancy due to adsorbed molecules.

The solution of Equations (5.1) and (5.2) yields:

$$B = W(1 - \frac{1}{A}) \qquad (5.3)$$

The number of adsorbed molecules, $N$, is

$$N = \frac{W}{M} N_{av} \qquad (5.4)$$

where $M$ = molecular weight of adsorbate and $N_{av}$ = Avogadro's number.

The average buoyancy, $b$, exerted on a single adsorbed molecule is derived from Equations (5.3) and (5.4) as:

$$b = \frac{B}{N} = \frac{(1 - 1/A)M}{N_{av}} \qquad (5.5)$$

The value of $b$ is constant if the adsorption is a monolayer type of coverage. Results from Figures 5.3 and 5.4 show that phenol/XAD-4 and PCP/XAD-4 are similar to monolayer adsorption systems.

If the large volume reservoir (150 liters) is replaced by a much smaller one (0.5–1.0 liter), then the adsorbed amount can be estimated by a mass balance in the liquid phase and compared with the gravimetric measurement. This batch mode calibration was performed and Equation (5.1) was applied to obtain the $A$-value for each adsorption system under study. Results showed that for each adsorption system the buoyancy factor $A$ was a constant, independent of adsorbed amount, as listed in Table 5.1.

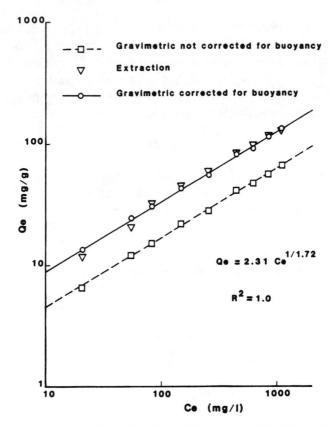

**Figure 5.3.** Isotherm and Buoyancy Effect for Phenol/XAD-4

Therefore, the adsorbed amount $Q$ can be obtained gravimetrically according to Equation (5.6).

$$Q = \frac{ARE}{W_s} \qquad (5.6)$$

Since the densities of the dilute aqueous solutions do not change significantly with solute concentration, the variation of $A$ at different liquid phase concentrations was negligible.

Theoretical Method. If the adsorbed molecules retain the original size that they had in the pure substance and displace water molecules of equal volume, the magnitude of $b$ can be calculated from the collision diameter of the adsorbate, $d_c$, as follows:

$$b = \frac{d_c^3 \pi \rho_w g}{6} \qquad (5.7)$$

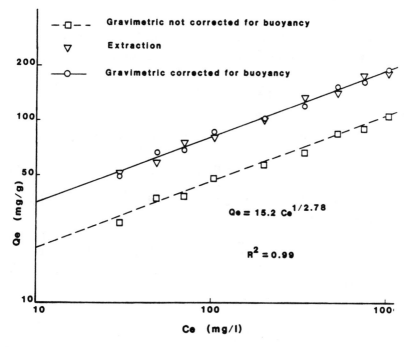

**Figure 5.4.** Isotherm and Buoyancy Effect for $p$-Chlorophenol/ XAD-4

**Table 5.1.** Values of Buoyancy Factor $A$

| Method | Phenol XAD-4 | PCP XAD-4 | Phenol F400 | PCP F400 |
|---|---|---|---|---|
| Measured with the apparatus developed here at batch mode | 2.0 | 1.7 | 8.5 | 4.1 |
| Determined by comparing the gravimetric values to the amount of adsorbate extracted | 1.9 | 1.82 | – | – |
| Theoretical prediction based on the assumptions stated in the text | 2.06 | 1.81 | 2.06 | 1.81 |

where $\rho_w$ = density of solution and $g$ = acceleration of gravity.

Equations (5.5) and (5.7) can be combined to give the value of $A$ as a function of physical parameters of the adsorbate only:

$$A = \frac{6M}{6M - N_{av}d_c^3\pi\rho_w g} \tag{5.8}$$

The value of $d_c$ can be determined from a correlation [9] as:

$$d_c = 0.833V_c^{1/3} \tag{5.9}$$

where $V_c$ = critical volume of adsorbate, which can be calculated by the Fedors' method [10].

The theoretical values of the buoyancy factor $A$ were calculated for both phenol and p-chlorophenol, and are listed in Table 5.1 for comparison. Results show that the agreement for the phenol/XAD4 and PCP/XAD4 is good, which implies the inert and hydrophobic surface of this synthetic resin. For adsorption onto F400, the theoretical buoyancy factors are much lower than the experimental values. Although a complete explanation can not be given at this stage, some mass substitution seems to take place in the carbon systems altering the force balance.

*Equilibrium Study*

When the gravimetric values $Q_g$ were correlated with solute concentration $C_e$, the Freundlich model could fit those equilibrium data:

$$\log Q_g = \log K_g + \frac{1}{n}\log C_e \tag{5.10}$$

where $Q_g = ER/W_s$.

As shown in Figures 5.3 and 5.4, however, these gravimetric values are smaller than the actual values because of the buoyancy effect. Since the buoyancy factor $A$ has been found to be independent of adsorbed amount, the actual equilibrium isotherm can be derived by introducing an $A$ value into Equation (5.10).

$$\log Q_e = \log A + \log K_g + \frac{1}{n}\log C_e \tag{5.11}$$

The real value of $K$ in the Freundlich isotherm is

$$K = AK_g \tag{5.12}$$

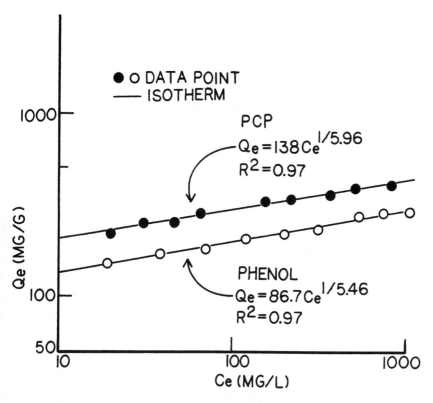

**Figure 5.5.** Isotherms for Phenol/F400 and $p$-Chlorophenol/F400

when the value of $A$ is known. The isotherms obtained are depicted in Figures 5.3 to 5.5, and Table 5.2 summarizes the correlated values of the equilibrium parameters for each adsorption system. The units used in this study are mg/l for $C_e$ and mg/g for $Q_e$.

*Kinetic Analysis*

The model adopted for the kinetic study is the solid diffusion model [11]:

$$\frac{\partial q}{\partial t} = \frac{D_s}{r^2} \frac{\partial}{\partial r} \left( r^2 \frac{\partial q}{\partial r} \right) \tag{5.13}$$

The experimental conditions lead to the following boundary and initial conditions:

$$\partial q / \partial t = 0 \qquad \text{at } r = 0 \tag{5.14}$$

$$q = Q_e \qquad \text{at } r = R_p, \text{ when } t > 0 \tag{5.15}$$

$$q = 0 \qquad \text{when } t = 0 \tag{5.16}$$

Table 5.2.    Equilibrium Parameters of Freundlich Isotherm

| Parameter | Phenol /XAD-4 | PCP /XAD-4 | Phenol /F400 | PCP /F400 |
|-----------|---------------|------------|--------------|-----------|
| $A$       | 2.06          | 1.81       | 8.5          | 4.1       |
| $K$       | 2.31          | 15.2       | 86.7         | 138.0     |
| $n$       | 1.72          | 2.78       | 5.46         | 5.96      |
| $R^2$     | 1.00          | 0.99       | 0.97         | 0.97      |

The analytical solution of these equations has been given [12] as:

$$\frac{Q_t}{Q_e} = 1 - \frac{6}{\pi^2} \sum_{m=1}^{\infty} \frac{1}{m^2} \exp\left(-\frac{\pi^2 D_s m^2 t}{R_p^2}\right) \qquad (5.17)$$

where $Q_t$ = average adsorbed amount at time $t$; $D_s$ = surface diffusivity and $R_p$ = particle radius.

Equation (5.17) can be applied by using gravimetric values directly, according to the following relation.

$$\frac{Q_t}{Q_e} = \frac{ARE_t/W_s}{ARE_e/W_s} = \frac{E_t}{E_e} \qquad (5.18)$$

where $E_t$ = extension of the spring at time $t$ and $E_e$ = extension of the spring at equilibrium. Therefore, the buoyancy effect can be excluded from the kinetic analysis. A typical set of adsorption rate data obtained from the gravimetric quartz spring apparatus is illustrated in Figure 5.6.

Since the normalized extension vs time profile is a function of $D_s$, a nonlinear regression procedure on the kinetic curve can provide the optimum value of $D_s$. The same analytic method has been reported for the kinetic study by using extraction in the differential reactor system [6,7]. It should be noted that the determination of $D_s$ this time is based only on the gravimetric measurements made during a single run. This means that the results are free of errors introduced when the data from two or more experiments are combined to provide adsorption rate information.

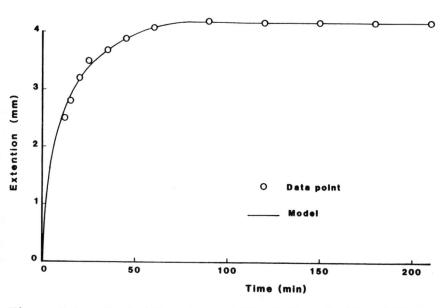

**Figure 5.6.**    Typical Experimental Kinetic Data for Phenol/XAD-4

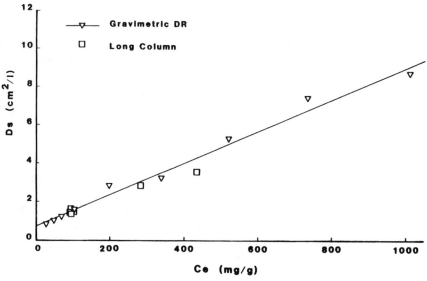

**Figure 5.7.**    Concentration Dependency of Diffusivity for
Phenol/XAD-4

The $D_s$ values obtained here showed a concentration dependency, as shown in Figures 5.7 to 5.10. The explanation of this $D_s$ variability can be found in the previous works [13-15] which provided

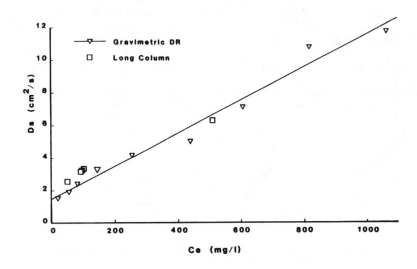

Figure 5.8.    Concentration Dependency of Diffusivity for
p-Chlorophenol/XAD-4

strong evidence that the surface diffusion coefficient changes with surface concentration. However, the consideration of this $D_s$ variation in adsorption modelling would require knowledge of the instantaneous local value of surface concentration (see Chapter III). This value cannot be measured directly at this time. Neither can it be predicted theoretically with the current state of knowledge concerning the internal geometry and structure of the adsorbent particle. Therefore, the surface diffusivity derived in this study is an effective value at that particular concentration.

In the same figures (Figures 5.7 to 5.10) the values of the intraparticle diffusion coefficients obtained from breakthrough data of a long column [16] are also plotted for comparison (see Chapter IX). The analysis of the column data employed the MADAM program [17,18]. The agreement between the values of the diffusion coefficient, obtained from the long column study and from the gravimetric method developed here was good.

Due to the complicated concentration variation in the adsorption zone, currently available computer models for long column design usually use constant rate parameters. Thus, for design purposes, the surface diffusivity in Equation (5.13) is still assumed to be a constant. The correlations between $D_s$ and $C_e$ developed here can be used to select a $D_s$ value appropriate for the design of a long column adsorber

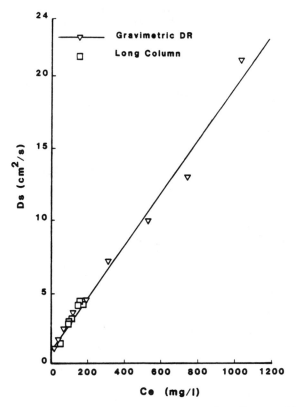

**Figure 5.9.**    Concentration Dependency of Diffusivity for Phenol/ F400

exposed to an influent concentration of $C_e$.

## CONCLUSIONS

The gravimetric quartz spring, used in air adsorption studies, has been successfully adapted to water systems. This was accomplished by identification and quantification of the buoyancy factor which, being negligible in air, is significant in water systems. The buoyancy factor was determined experimentally and found to be constant for each particular adsorption system.

It was possible to explain the buoyancy effect theoretically, in terms of equivolume displacement of water, for the XAD4 systems, but not for those of carbon. In the carbon systems the buoyancy factor was three times larger than predicted by the equivolume displacement assumption.

The similarity of the rate parameters measured by the differential

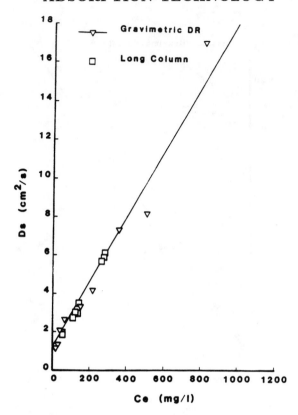

**Figure 5.10.** Concentration Dependency of Diffusivity for
*p*-Chlorophenol/F400

reactor at constant solute concentration to those from the long column
study suggests that the type of parameters obtained in this study can
be applied to column design.

The spring balance configuration, combining the mathematical advantages of a differential reactor and the simplicity of the gravimetric
measurement, was shown to be an effective tool for the measurement
of equilibrium and kinetic parameters of adsorption.

*REFERENCES*

1. McBain, J.W. *The Sorption of Gases and Vapors by Solid.* George Routledge & Sons, London, England (1934).
2. Grant, R.J., M. Manes, and S.B. Smith, *AIChE. J.*, **8**, 403 (1962).
3. Nakahara, T., M. Hirata, and S. Komatsu, *J. Chem. Eng. Data*, **26**, 161 (1981).
4. Nakahara, T., M. Hirata, and H. Mori, *J. Chem. Eng. Data*, **27**, 317 (1982).
5. Urano, K., S. Omori, and E. Yamamoto, *Envir. Sci. Tech.*, **16**, 10 (1982).
6. Noll, K.E., A.A. Aguwa, Y.P. Fang, and P.T. Boulanger, *ASCE. EE.*, **111**, 487 (1985).
7. Aguwa, A.A. *Evaluation of Sorption Behavior of Organics with Synthetic Resin.* Ph.D. Thesis, Illinois Institute of Technology, Chicago, IL (1984).
8. Aguwa, A.A., J.W. Patterson, C.N. Haas, and K.E. Noll, *J. WPCF.*, **56**, 442 (1984).
9. Hirschfelder, J.O., R.B. Bird, and E.L. Spotz, *Chem. Rev.* 44, 205 (1949).
10. Fedors, R.F. *AIChE. J.*, **25**, 202 (1979).
11. Rosen, J.B. *J. Chem. Phys.*, **20**, 387 (1952).
12. Crank, J. *The Mathematics of Diffusion.* Oxford University Press, London, England (1956).
13. Gilliland, E.R., R.F. Baddour, G.P. Perkinson, and K.J. Sladek, *Ind. Eng. Chem. Fundam.*, **13**, 95 (1974).
14. Higashi, K., H. Ito, and J. Oishi, *J. Atomic Eng. Soc. Japan*, **5**, 846 (1963).
15. Suzuki, M., and T. Fujii, *AIChE. J.*, **28**, 380 (1982).
16. Hou, W.S. *Modeling Analyses of Fixed Bed Adsorbers in Liquid-Solid Systems.* Ph.D. Thesis, Illinois Institute of Technology, Chicago, IL (1989).
17. Crittenden, J.C., and W.J. Weber, Jr., *ASCE. EE.*, **104**, 185 (1978).
18. Crittenden, J.C., and W.J. Weber, Jr., *ASCE. EE.*, **104**, 433 (1978).

# CHAPTER VI

---

## COMBINED DIFFUSION
## DIFFERENTIAL REACTOR MODELS

---

Contributing Authors: C. Arai and M.C. Yeh

## INTRODUCTION

In the treatment of the diffusion phenomena in porous materials it is often necessary to consider pore diffusion together with surface diffusion. It is well known that the pore diffusion coefficient can be obtained from correlations that use the molecular diffusion coefficient, the porosity, and the tortuosity factor of the porous materials. The molecular diffusion coefficient is usually calculated by applying correlated equations (see Chapter II), and the porosity can be determined experimentally by the nitrogen desorption method or mercury porosimeter. However, the value of particle tortuosity is not always available, and is not easily assessed by experimental method. This chapter presents two methods of approach to this problem, through a combined diffusion model which was solved numerically by an orthogonal collocation method.

## LIQUID-SOLID SYSTEM

The tortuosity factor is a measure of the actual length an adsorbing molecule must travel, as compared to the radial distance travelled, and is important in defining the construction of porous material. For the determination of tortuosity through liquid-solid systems, the proposed methods and models include the combined diffusion model with the batch reactor method [1-3], the pore diffusion model with the batch reactor method [4-6], and the pore diffusion model with the Wicke-Kallenbach method [7]. Those studies showed that the estimation of the tortuosity could affect the model significantly if the pore diffusion was a rate control factor during the adsorption process.

### Method and Results

Adsorption rate data collected from a differential reactor in liquid-solid systems (see Chapter III and V) can be used to evaluate simultaneously both the tortuosity factor and the surface diffusion coefficient. In order to derive the fundamental equation, it is assumed that the incremental increase of the amount adsorbed is due to both surface and pore diffusion, and that the accumulation term in the pore water can be neglected  because the adsorbed phase concentration is much larger than that in the water. Equation (2.50) is used to describe the instantaneous mass balance in a spherical particle. It is also assumed that no resistance is due to the adsorption-desorption process at the active sites in a pore. The relationship of local equilibrium at the interphase is related by a Freundlich-type isotherm.

The pore diffusion coefficient in a liquid-solid system does not include the Knudsen effect, therefore,

$$D_p = \epsilon_p D_m / \tau \qquad (6.1)$$

Thus, Equation (2.50) contains four parameters: surface diffusion coefficient, particle density,  porosity, and tortuosity. Since density and porosity can be measured without the application of the adsorption technique, as a result, Equation (2.50) contains two unknown parameters. Therefore, when values are given for these parameters, it is possible to solve Equation (2.50). In the following analysis, the numerical procedure is developed by using the orthogonal collocation method [8,9] to solve Equation (2.50), based on the initial conditions:

$$C_p = 0, \qquad q = 0 \qquad \text{at } t = 0 \qquad (6.2)$$

and on the boundary condition:

$$\rho_p \frac{\partial}{\partial t} \int_0^{R_p} q r^2 \, dr = k_f R_p^2 (C_0 - C_s) \qquad (6.3)$$

By using the numerical solution of Equation (2.50), the adsorbent amount, $Q_t$, at time $t$ is given by:

$$Q_t = \frac{3}{R_p^3} \int_0^t \frac{\partial}{\partial t} \int_0^{R_p} q r^2 \, dr dt \qquad (6.4)$$

The integration for $r$ in Equation (6.4) can be carried out with the aid of the weighting value of the orthogonal collocation method, and for $t$, numerically.

By defining the average difference, $\delta$, between $(Q_t)_n$, numerical solution, and $(Q_t)_e$, experimental data, an equation is given as:

$$\delta = \frac{\sum_{i=1}^N |(Q_t/Q_e)_n - (Q_t/Q_e)_e|}{N} \qquad (6.5)$$

The tortuosity and surface diffusion coefficient can be determined such that $\delta$ is minimized.

If the equilibrium isotherm is a linear function, $q = KC$, then Equation (2.50) can be simplified as:

$$\frac{\partial q}{\partial t} = \left(D_s + \frac{\epsilon_p D_m}{K \rho_p \tau}\right) \frac{1}{r^2} \frac{\partial}{\partial r} \left(r^2 \frac{\partial q}{\partial r}\right) \tag{6.6}$$

In this case, even if $D_s$ and $\tau$ have different values, the solution of Equation (6.6) is identical when $(D_s + \epsilon_p D_m / K \rho_p \tau)$ has the same value. Therefore, there are infinite combinations of $D_s$ and $\tau$ for which $\delta$ in Equation (6.5) becomes a minimum. But the Freundlich coefficient, $n$, is not generally equal to unity in the range of ordinary concentrations, and therefore, a unique solution is possible.

Figure 6.1 shows the variations of $\delta$ calculated by using Equation (6.5) ($\tau = 3.0$ and $D_s = 2.0 \times 10^{-8}$ cm$^2$/sec are assumed for the kinetic data). It is clear that $\delta$ achieves a minimum point at the given point **P**, and there is only one minimum point. The fact that the minimum value can be obtained is the proof of availability on the present method. The data used in this study were collected with the differential reactor for the liquid-solid system (Chapter III). Figure 6.2 depicts typical kinetic data for $p$-chlorophenol in the XAD-4 system.

Figure 6.3 shows the tortuosity factors of XAD-2 and XAD-4. It is clear that points in the figure are independent of influent concentration and the adsorbates. All the points fall in the range of 2.6 to 3.4 for XAD-4, and 3.1 to 3.7 for XAD-2. The average values are 3.1 and 3.3, as shown by the dotted lines. The average tortuosity value, their standard deviations and tortuosity factor obtained from other sources [2,3] are given in Table 6.1. Although there are differences between the values in this study and the ones from the literature, it can not be determined whether these differences are due to experimental error or due to experimental methods (differential vs. batch reactor).

Tortuosity is considered a strong function of the pore alignment and the varying pore cross sectional area. If a random pore structure is assumed, it can be shown that the tortuosity should be a constant and equal to 3.0. Values greater than 3.0 are attributed to a structural heterogeneity such as skin effect, dead end pores, and anisotropy, while values less than 3.0 may be attributed to large short pores which allow the transport paths to cut corners. With the concept of a factor of 1/3 in every orientation for a random pore structure, a network permeability model suggests that the tortuosity is 3.0 for porous media [10].

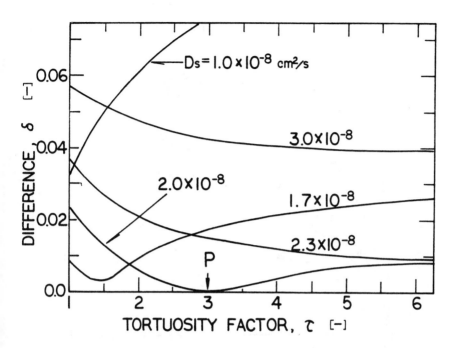

$-D_s = 1.0 \times 10^{-8}$ cm²/s

$3.0 \times 10^{-8}$

$2.0 \times 10^{-8}$

$1.7 \times 10^{-8}$

$2.3 \times 10^{-8}$

Figure 6.1.    The Variation of Objective Function ($n = 2.5$)

Satterfield and Cadle [11] studied 17 commercially manufactured pelleted catalysts and found that most tortuosity factors fell between 3.0 and 7.0; for about half the catalysts this value is about 4.0 regardless of macroporosity or composition. The average value is about 3.6 for mesoporous pellets and 2.8 for macroporous particles. Other reported values of tortuosity are listed as follows: 4.0 for benzene/F200 [1]; 3.9 for benzaldehyde/XAD1 [3]; 1.7 for 2-dodecylbenzene sulfonate (DBS)/CAL, 2.65 for DBS/Takeda X4401, and 6.2 for DBS/Takeda 911 [4]; 2.8 for benzaldehyde/XAD-4 [5]; 3.1 for $O_2$/BPL and 3.2 for $H_2SO_4$/BPL [6]; 3.22–3.48 for NaCl or KCl/carbon RGW488 [7].

Figure 6.4 shows the surface diffusion coefficients determined using the tortuosity from Figure 6.3 in the combined diffusion model (Equation (2.50)). Results show that the coefficients are dependent upon the influent concentration. Compared with the kinetic analysis in Chapter III and V, it is evident that the concentration-dependent character of the effective diffusion coefficient results from the relationship between the surface diffusivity and concentration.

By utilizing both the tortuosity and diffusion coefficient, the best-fit curves were generated, as shown by the solid lines in Figure 6.2. Those curves show good agreement with the experimental data. In

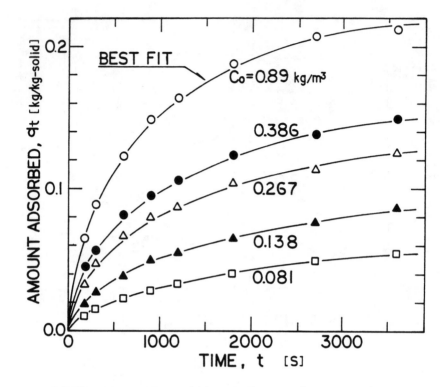

**Figure 6.2.**    Typical Kinetic Data and Best Fit Curves

this process, the total mass transfer rate $(\partial q/\partial t)$, and the contribution of the pore diffusion rate (PDR), the second term in the right hand side of Equation (2.50), to the total diffusion rate (TDR) were also calculated, since the tortuosity factor is determined by the contribution of the pore diffusion phenomenon. Figure 6.5 depicts the results. The mass transfer rate increases with increasing liquid concentration, because the surface diffusion is concentration dependent. The ratio of PDR/TDR at initial moment also increases with concentration, and its average value for an entire process is about 0.2.

*Sensitivity Analysis*

In order to determine the effects of measurement error, the tortuosity factors and diffusion coefficients were calculated by selecting three runs from Figure 6.2 as examples.

Figure 6.6 shows the effects of error in particle radius. A relative error of 10% on the horizontal axis, corresponds to 0.225 mm. As shown in the figure, the error in tortuosity changes from 2 to 4%, for

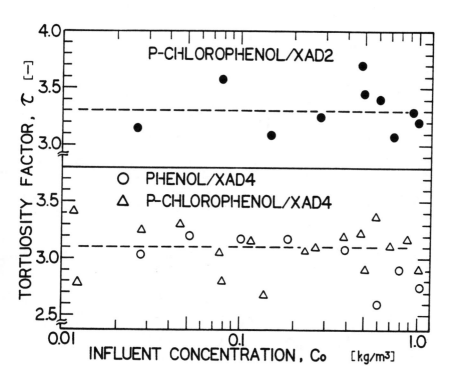

Figure 6.3.    Tortuosity Factors for XAD-2 and XAD-4

Table 6.1.    Tortuosity Factors

| Sorbent | Sorbate | $\tau$ | Data No. | Std. Dev. | Reference |
|---------|---------|--------|----------|-----------|-----------|
| XAD-2 | $p$-Chlorophenol | 3.3 | 10 | 0.22 | 3.4 [3] |
| XAD-4 | Phenol $p$-Chlorophenol | 3.1 | 25 | 0.21 | 2.7 [2], 3.1 [3] |

$\tau$ = Tortuosity
Std. Dev. = Standard Deviation

the 10% change in particle radius, but the error in tortuosity increases rapidly with increasing particle radius. The error curves for the diffusion coefficient shows that errors of about 30% are possible for a 10%

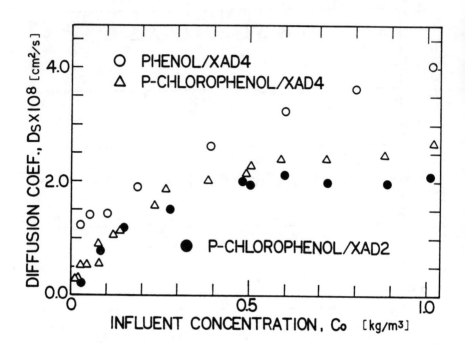

Figure 6.4.   Surface Diffusion Coefficients versus Influent
    Concentration

error in particle radius. This suggests that measurements of particle radius should be accurately made. In this study, particle radius was determined by using a microscope, as mentioned in Chapter III.

Figure 6.7 depicts the effect of variations in the external mass transfer coefficient, $k_f$. The $k_f$ values are usually estimated by using a correlation equation. In this study, Equation (2.25) [12] was used after comparison with other correlations [13]. Figure 6.7 shows that errors in the mass transfer coefficient have a smaller effect on both the tortuosity and diffusion coefficient than errors in particle radius. However, the relationship in the figure is dependent on the magnitude of the mass transfer coefficient. The solid triangle points represent half values of the original mass transfer coefficient.

Figure 6.8 shows the effect of errors in measuring the total amount adsorbed at equilibrium. In this case, the solution of Equation (2.50) does not agree with the experimental data. Therefore, in order to eliminate problems due to isotherm errors, it is important to determine if the solution of Equation (2.50) is the best-fit curve to the original

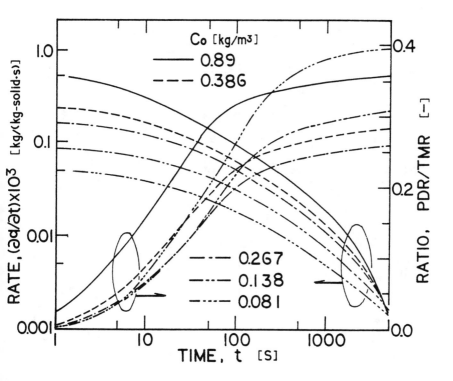

Figure 6.5.    Adsorption Rates and the Ratio of Pore Diffusion
    Rate (PDR) to the Total Mass Transfer Rate (TDR)

data.

## GAS-SOLID SYSTEMS

The pore diffusion model is generally considered in gas-solid adsorption systems  because of the Knudsen diffusion effect. Knudsen diffusion is encountered in pores containing a gas  when the pressure is low or the pore size is small. If the mean free path of the gas molecules is large compared to the pore diameter, then the molecules collide much more frequently with the pore walls than with one another. The resistance to diffusion along the pores is then due primarily to molecular collisions with the wall rather than with one another, as in ordinary diffusion. In some ranges of gas densities and pore sizes both types of collision are important.

To estimate the pore diffusion coefficient, the value of particle tortuosity has to be determined for each kind of adsorbent. The available methods  and models in the literature are the pore diffusion model with chro-

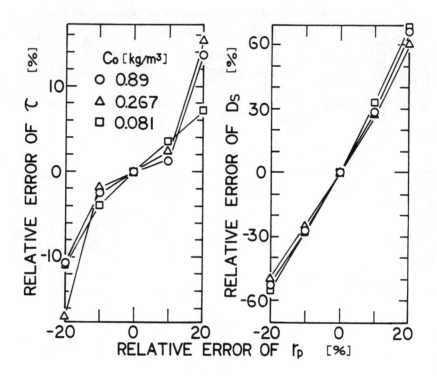

Figure 6.6.    Effect of Error in Particle Size

matography method [14–17], the pore diffusion model with the Wicke-Kallenbach method [18], and the combined diffusion model with the transient response method [19].

In this section, the $D_s$ and $D_p$ of air adsorption systems were obtained by analyzing the differential reactor rate data according to the method presented for the water systems. The rate data presented in Chapter IV analyzed by the surface diffusion model are used here to demonstrate the method. Table 6.2 lists the adsorption systems and concentration ranges analyzed in this study.

### $D_s$, $D_p$ Determination

The combined diffusion model was used to simulate the loading history of an adsorbent particle and to estimate the relative importance of surface and pore fluxes. The model assumes that the external mass transfer resistance can be neglected [20–22] and that surface diffusivity is constant for each run. Then Equation (2.46) can be expressed

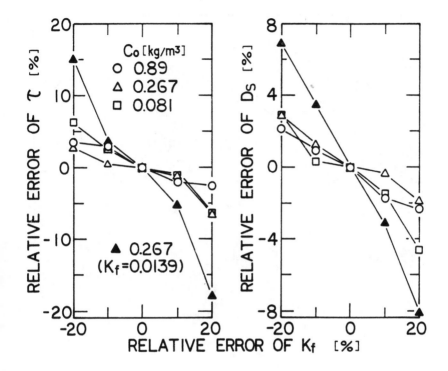

**Figure 6.7.**    Effect of Error in Mass Transfer Coefficient

in the following dimensionless form:

$$\left(1 + \frac{A}{n}\overline{C}_p^{1/n-1}\right)\frac{\partial \overline{C}_p}{\partial \overline{T}} = \left(1 + \overline{D}\frac{A}{n}\overline{C}_p^{1/n-1}\right)\frac{1}{\overline{r}^2}\frac{\partial}{\partial \overline{r}}\left(\overline{r}^2\frac{\partial \overline{C}_p}{\partial \overline{r}}\right)$$
$$+ \overline{D}\frac{A}{n}\left(\frac{1}{n} - 1\right)\overline{C}_p^{1/n-2}\left(\frac{\partial \overline{C}_p}{\partial \overline{r}}\right)^2 \tag{6.7}$$

The boundary and initial conditions were simplified as:

$$\overline{C}_p = \overline{C}_{p0}, \quad \overline{q} = \overline{q}_0 \qquad \text{when } \overline{T} = 0 \tag{6.8}$$

$$\partial \overline{C}_p/\partial \overline{T} = 0, \quad \partial \overline{q}/\partial \overline{r} = 0 \qquad \text{at } \overline{r} = 0 \tag{6.9}$$

$$\overline{C}_p = 1.0, \quad \overline{q} = 1.0 \qquad \text{at } \overline{r} = 1 \tag{6.10}$$

where

$$\overline{T} = \frac{tD_p}{R_p^2} \tag{6.11}$$

$$A = \frac{\rho_p K C_0^{1/n-1}}{\epsilon_p} = \frac{\rho_p Q_e}{\epsilon_p C_0} \tag{6.12}$$

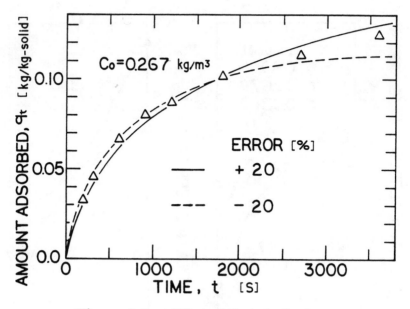

**Figure 6.8.**    Effect of Error in Isotherm

$$\overline{r} = \frac{r}{R_p} \qquad (6.13)$$

$$\overline{D} = \frac{D_s}{D_p} \qquad (6.14)$$

$$\overline{q} = \frac{q}{Q_e}, \quad \overline{q}_0 = \frac{q_0}{Q_e} \qquad (6.15)$$

$$\overline{C}_p = \frac{C_p}{C_0}, \quad \overline{C}_{p0} = \frac{C_{p0}}{C_0} \qquad (6.16)$$

At any time interval $\overline{T}$, the average uptake is:

$$\overline{Q}_T = \frac{Q_t}{Q_e} = 3 \int_0^1 \overline{q}\,\overline{r}^2\,d\overline{r} \qquad (6.17)$$

The solution to Equation (6.7) utilizes the orthogonal collocation method. Using the same Laplacian operators and summation formula for the function $\overline{C}_p(\overline{r}_i)$ as in the long column model (see Chapter IX),

Table 6.2.    Contribution of Surface Diffusion to the Total Mass
Transfer Process

| System | $C_0$ Range (ppm) | $D_p$ Range $\times 10^{-3}$ (cm²/sec) | $D_s$ Range $\times 10^{-3}$ (cm²/sec) | $R_D$ (–) |
|---|---|---|---|---|
| $C_6H_6$/XAD-4 | 85–8437 | 0.91–1.40 | 2.35–35.5 | 0.67–0.56 |
| $C_6H_6$/BAC | 85–8437 | 1.21–1.65 | 1.30–72.2 | 0.94–0.88 |
| $C_7H_8$/XAD-4 | 350–7000 | 0.95–1.22 | 5.60–31.0 | 0.72–0.80 |
| $C_7H_8$/XAD-4 | 70–7000 | 1.01–1.41 | 1.74–103.0 | 0.99–0.85 |
| $p$-$C_8H_{10}$/XAD-4 | 60–6046 | 0.88–1.20 | 1.71–18.5 | 0.85–0.72 |
| $p$-$C_8H_{10}$/BAC | 60–6046 | 1.09–1.58 | 1.02–85.8 | 0.92–0.79 |
| $CCl_4$/XAD-4 | 77–7665 | 0.82–1.02 | 6.90–74.1 | 0.78–0.65 |
| $CCl_4$/BAC | 77–7665 | 0.89–1.10 | 2.85–83.6 | 0.95–0.90 |
| $C_2HCl_3$/XAD-4 | 84–8277 | 0.96–0.98 | 4.25–55.7 | 0.80–0.73 |
| $C_2HCl_3$/BAC | 84–8277 | 0.93–1.12 | 1.35–68.4 | 0.95–0.84 |
| $C_2Cl_4$/XAD-4 | 73–7274 | 0.77–0.92 | 1.96–16.7 | 0.83–0.75 |
| $C_2Cl_4$/BAC | 73–7275 | 0.87–1.10 | 1.04–52.9 | 0.94–0.80 |

Equation (6.7) becomes:

$$\frac{\partial \overline{C}_p}{\partial \overline{T}}\Big|_{\overline{r}_i} = \left(1 + \frac{A}{n}\overline{C}_{P(\overline{r}_i)}^{1/n-1}\right)^{-1}\left[\left(1 + \overline{D}\frac{A}{n}\overline{C}_{P(\overline{r}_i)}^{1/n-1}\right)\sum_{j=1}^{Nc} B_{ij}\overline{C}_p(\overline{r}_j)\right.$$

$$\left. + \overline{D}\frac{A}{n}(\frac{1}{n} - 1)\overline{C}_{P(\overline{r}_i)}^{1/n-2}(\sum_{j=1}^{Nc} A_{ij}\overline{C}_p(\overline{r}_j))^2\right]$$

(6.18)

The selected orthogonal polynomial has a symmetric form so the boundary codition at $\overline{r} = 0$ is implicitly included in the collocation coefficients. The boundary condition at the particle surface reduced $\overline{C}_p(\overline{r}_{Nc})$ in Equation (6.18) to 1.0. The system of these equations was solved using the Gears stiff variable step integration algorithm with full Jacobin analysis.

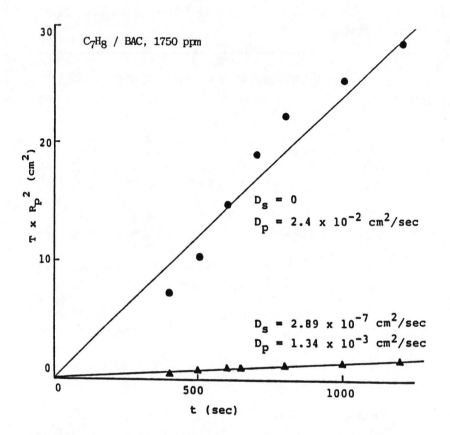

**Figure 6.9.** The Relationship between Experimental $t$ and Calculated $T \times R_p^2$ with Parameter $D_s/D_p$

Equation (6.7), whose boundary and initial conditions were simplified in Equations (6.8)–(6.10), was used to describe the rate of mass transfer within the adsorbent particle and to determine the intraparticle diffusion coefficients. The solution of these equations yields a curve of normalized uptake $\overline{Q}$ versus dimensionless time $\overline{T}$ for a given value of $D_s/D_p$. The predicted curve can then be compared with the experimental curve of $\overline{Q}$ versus $\overline{T}$, which corresponds to the same time $t$. The definition of $\overline{T}$ suggests that if $\overline{T}$ is plotted versus $t$, the line should pass exactly through the given point (0,0) and with a slope of $D_p/R_p$. An example is illustrated in Figure 6.9.

The slope of the least-squares line through a given point (0,0) can be calculated using Equation (6.19) [23]:

$$b_s = \frac{\sum t_i \overline{T}}{\sum t_i^2}$$ (6.19)

where $b_s$ = slope of the least-squares line, $t_i$ = various experimental time.

In order to get the best possible fit, $D_s/D_p$ is varied in the range of 0.0–0.1. A curve-fitting procedure is used to minimize the standard deviation of the least-squares line. Consequently, an objective function, $\epsilon$, was defined as:

$$\epsilon = \sum d_i^2 = \sum (\overline{T}_i - b_s t_i)^2$$ (6.20)

where $d_i$ = the vertical distance, parallel to the $\overline{T}$-axis of $(t_i, \overline{T}_i)$ from the line.

The minimum of the objective function defines the value of $D_s/D_p$. This method provides an optimum $D_s/D_p$ and optimum pore diffusivity $D_p$. An example in Figure 6.9 shows that $D_s = 2.89 \times 10^{-7}$ cm$^2$/sec and $D_p = 1.34 \times 10^{-3}$ cm$^2$/sec are the optimum values for analysis. The pore and surface diffusion coefficients of benzene, toluene, p-xylene, carbon tetrachloride, trichloroethylene and tetrachloroethylene on XAD-4 and activated carbon at various vapor concentrations and at 25$^0$C were measured in this way.

These measurements indicated that the $D_p$ value of each particular sorbent-sorbate system was, as expected, independent of concentration. The average $D_p$ values at 25$^0$C are given in Table 6.3. In contrast to $D_p$, a strong increase of $D_s$ with solid phase concentration was exhibited by all the adsorbent-adsorbate systems studied. Figures 6.10 and 6.11 show the $D_s$ values for all systems plotted versus the vapor concentration.

The relative contribution of surface diffusion to intraparticle mass transfer was estimated as the ratio:

$$R_D = D_s / \left[ D_s + \frac{\epsilon_p D_p}{\rho_p} \overline{\frac{\partial C_p}{\partial q}} \right]$$ (6.21)

The values of $R_D$ are found in the range of 60–80% for XAD-4, and 80–99% for activated carbon. The actual contributions of $D_s$ and $D_p$ to the total mass transfer are shown in Table 6.2 for each system.

The $D_p$ values so obtained can be used to estimate the tortuosity factors of the adsorbents by using Equations (2.33)–(2.38). The

Table 6.3. $D_p$ Values* at $25^0$C and $\tau$ Values*

| Gas | Sorbent | $D_p \times 10^3$ (cm²/sec) | Data No. (–) | $\tau$ (–) | S.D. (–) |
|---|---|---|---|---|---|
| CCl₄ | BAC | 0.987 | 7 | 4.2 | 0.32 |
| C₂HCl₃ | BAC | 1.01 | 7 | 4.4 | 0.31 |
| C₂Cl₄ | BAC | 0.993 | 7 | 4.0 | 0.41 |
| C₆H₆ | BAC | 1.44 | 7 | 4.0 | 0.43 |
| C₇H₈ | BAC | 1.25 | 7 | 4.2 | 0.51 |
| p-C₈H₁₀ | BAC | 1.29 | 7 | 3.8 | 0.48 |
| Average | | - - - | 42 | 4.1 | 0.35 |
| CCl₄ | XAD-4 | 0.9 | 7 | 2.9 | 0.22 |
| C₂HCl₃ | XAD-4 | 0.97 | 7 | 2.9 | 0.02 |
| C₂Cl₄ | XAD-4 | 0.81 | 7 | 3.1 | 0.24 |
| C₆H₆ | XAD-4 | 1.07 | 7 | 3.4 | 0.47 |
| C₇H₈ | XAD-4 | 1.11 | 5 | 3.0 | 0.30 |
| p-C₈H₁₀ | XAD-4 | 1.01 | 7 | 3.1 | 0.35 |
| Average | | - - - | 40 | 3.1 | 0.27 |

\* Average value based on data points for each system.
S.D.= Standard Deviation.

average values of these tortuosities were 3.1 for XAD-4 and 4.1 for carbon. The particular values measured are shown in Table 6.3 and in Figure 6.12. Other tortuosities reported for different systems are listed as follows: 3.2–9.5 for He or $O_2$/Boehmite-Al [14]; 5.3–6.1 for benzene/Davison-525 and 1.6–2.3 for benzene/Linde-5A [15]; 3.2 for $N_2$/Zeolite-5A [16]; 4.5 for $CO_2$/BPL [17]; 3.34 for He/Boehmite-Al [18]; 4.0 for propylene/Linde-13X [19]. Compared with those results, the tortuosity values estimated using the combined diffusion model and differential reactor method are very reasonable.

## CONCLUSIONS

A method for determining the tortuosity factor and surface diffusion coefficient is presented in this chapter. Because the external concen-

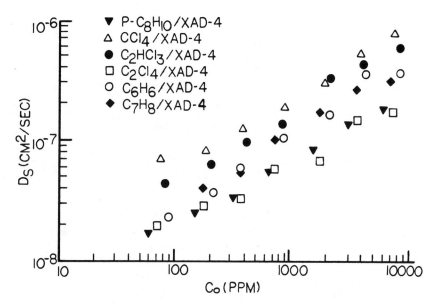

**Figure 6.10.** Surface Diffusion Coefficient versus Vapor Concentration (XAD-4)

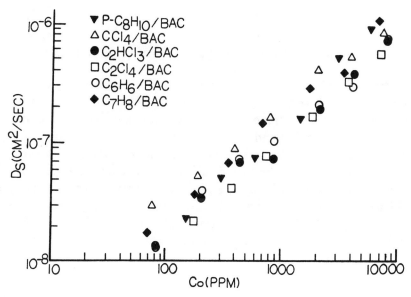

**Figure 6.11.** Surface Diffusion Coefficient versus Vapor Concentration (BAC)

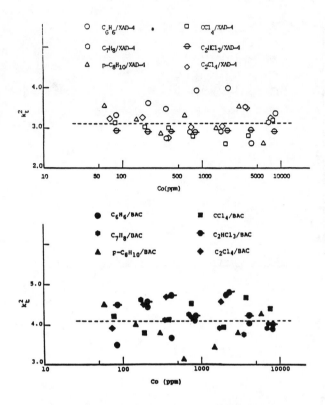

Figure 6.12.     Tortuosity Factor for XAD-4 and BAC

tration was kept constant, and the structure of adsorbents was close
to uniform, the possible errors from those factors were reduced to a
minimum. Since the tortuosity is a structural parameter, it is independent of temperature and the nature of the diffusing species [15,16].
The results show that the tortuosity is independent of diffusing species
and concentration, and the surface diffusion coefficient is a function of
external concentration. The same tendency can be observed in both
liquid-solid and gas-solid systems, although the methods of analysis
are different. The agreement between the tortuosity values for XAD-4
obtained from the air and water systems was excellent.

*REFERENCES*

1. Furusawa, T., and J.M. Smith, *AIChE. J.*, **19**, 401 (1973).
2. Komiyama, H., and J.M. Smith, *AIChE. J.*, **20**, 728 (1974).
3. Komiyama, H., and J.M. Smith, *AIChE. J.*, **20**, 1110 (1974).
4. Suzuki, M., and K. Kawazoe, *J. Chem. Eng. Japan*, **7**, 346 (1974).
5. Furusawa, T., and J.M. Smith, *AIChE. J.*, **20**, 88 (1974).
6. Komiyama, H., and J.M. Smith, *AIChE. J.*, **21**, 664 (1975).
7. Shibuya, H., and Y. Uraguchi, *J. Chem. Eng. Japan*, **10**, 446 (1977).
8. Villadsen, J.V., and W.E. Stewart, *Chem. Eng. Sci.*, **22**, 1483 (1967).
9. Finlayson, B.A. *Chem. Eng. Sci.*, **26**, 1081 (1971).
10. Dullien, F.A.L. *AIChE. J.*, **21**, 299 & 820 (1975).
11. Satterfield, C.N., and P.J. Cadle, *Ind. and Eng. Chem. Proc. Des. Dev.*, **7**, 256 (1968).
12. Wakao, N., and T. Funazkri, *Chem. Eng. Sci.*, **33**, 1375 (1978).
13. Foo, S.C., and R.G. Rice, *AIChE. J.*, **21**, 1149, (1975).
14. Suzuki, M., and J.M. Smith, *AIChE. J.*, **18**, 326 (1972).
15. Lee, L.K., and D.M. Ruthven, *Ind. and Eng. Chem. Fund.*, **16**, 290 (1977).
16. Hashimoto, N., and J.M. Smith, *Ind. and Eng. Chem. Fund.*, **12**, 353 (1973).
17. Andrieu, J., and J.M. Smith, *AIChE. J.*, **26**, 944 (1980).
18. Doğu, G., and J.M. Smith, *AIChE. J.*, **21**, 58 (1975).
19. Kelly, J.F., and O.M. Fuller, *Ind. and Eng. Chem. Fund.*, **19**, 11 (1980).
20. Noll, K.E., C.N. Haas, A.A. Aguwa, M. Satoh, A. Belalia, and P.S. Bartolomew, *Direct Differential Reactor Studies on Adsorption from Industrial Strength Liquid and Gaseous Solutions.* Presentation at the Engineering Foundation Conference at Schloss Elmau, Bavaria, West Germany (1983).
21. Noll, K.E., and J.N. Sarlis, *J. APCA.*, **38**, 1512 (1988).
22. Carlson, N.W. and J.S. Dranoff, *Ind. Chem. Process Des. Dev.*, **24**, 1300 (1985).
23. Prahl, W.H. *Chem. Eng.*, **5**, 65 (1983).

# CHAPTER VII

## KINETIC STUDIES OF
## MULTICOMPONENT ADSORPTION
## USING DIFFERENTIAL REACTOR TECHNIQUE

### INTRODUCTION

Multicomponent adsorption kinetics can also be evaluated using the differential reactor technique [1-3]. Adsorption systems evaluated in this chapter include two binary mixtures (phenol+$p$-chlorophenol and phenol+2,5-dichlorophenol) and one ternary mixture (phenol+$p$-chlorophenol+2,5-dichlorophenol) adsorbed on synthetic resin XAD-4. The kinetic data reported in this study are based on solid phase adsorption and provide a unique set of data. To interpret the competitive and interactive phenomena appearing in the kinetic data, two intra-particle diffusion models were selected for comparison, one is a solid diffusion model (Equation (2.39), and the other is a combined diffusion model (Equation (2.50)). The isotherm model used for describing local equilibrium is the IAS model.

Mathews and Su [4] obtained good predictions by using a homogeneous surface diffusion model in a binary kinetic study. Fritz et al. [5] compared several models in a binary kinetic study and found that the film-heterogeneous diffusion model usually performed best. Liapis and Rippin [6] compared different models and emphasized that pore diffusion is important in the multicomponent adsorption systems. However, all of their kinetic studies were conducted in a batch reactor, and the results are based on liquid phase measurements.

Although diffusional interaction has been proposed in binary adsorption systems [7,8], independent diffusivities were assumed in this study. Therefore, both models can be extended to the case of ternary adsorption systems by using the information derived from single component systems.

The applied numerical method is the orthogonal collocation method [9,10]. This method has proved useful in solving highly non-linear differential equations [6,11].

### ANALYTICAL RESULTS AND DISCUSSION

*Adsorption Equilibrium*

The IAS model, Equations (2.15)–(2.20) (Chapter II), with the Freundlich isotherm was used to evaluate the data. Parameters for the single component isotherms are listed in Table 7.1.

**Table 7.1.**   Values of Parameters* in Isotherms

| Parameter | Phenol/XAD-4 | PCP/XAD-4 | 2,5-DCP/XAD-4 |
|---|---|---|---|
| Freundlich Isotherm | | | |
| $K$ | 2.01 | 12.96 | 41.24 |
| $n$ | 1.734 | 2.353 | 2.683 |
| $R^2$ | 0.99 | 0.99 | 0.99 |
| Three-Parameter Isotherm | | | |
| $K$ | 0.9113 | 7.0847 | 118.59 |
| $n$ | 0.1473 | 0.2538 | 2.5890 |
| $\gamma$ | 0.575 | 0.691 | 0.643 |
| Ref. | [1] | [1] | [2] |

* Values were evaluated for $C$[mg/l] and $q$[mg/g].

Multicomponent adsorption equilibria were carried out in batch reactors [1-3]. Experimental data were correlated empirically by using a modified three-parameter model [12] as follows:

$$q_i = \frac{A_i(C_i/\eta_i)}{1 + \sum_{j=1}^{K} B_j(C_j/\eta_j)^{\gamma_j}} \tag{7.1}$$

The interaction parameter, $\eta_i$, estimated from multicomponent equilibrium studies is listed in Table 7.2. Statistical results of the multicomponent adsorption equilibrium models are listed in Table 7.3.

*Modeling Comparison*

Surface diffusion coefficients using the combined diffusion models are shown in Figure 7.1. The pore diffusion coefficient for the combined diffusion model was estimated as $D_p = D_m \epsilon_p / \tau$, where particle

Table 7.2.    Values of $\eta^*$ in Binary and Ternary Systems

| System | Phenol | PCP | 2,5-DCP | Ref. |
|---|---|---|---|---|
| Phenol+PCP/ XAD-4 | 0.719 | 0.764 | | [1] |
| Phenol+2,5-DCP/ XAD-4 | 0.25 | | 1.12 | [2] |
| Phenol+PCP+2,5-DCP/ XAD-4 | 0.2004 | 0.2695 | 0.8535 | [3] |

* Values were evaluated for $C$[mg/l] and $q$[mg/g].

tortuosity was assumed to be 3.1, according to theoretical considerations [13,14] and experimental results (Chapter VI). The values of
pore diffusion coefficients are $1.72 \times 10^{-6}$ cm$^2$/sec for phenol/XAD-
4, $1.56 \times 10^{-6}$ cm$^2$/sec for $p$-chlorophenol/XAD-4, and $1.29 \times 10^{-6}$
cm$^2$/sec for 2,5-dichlorophenol/XAD-4. Compared with the effective
diffusion model in Chapter III, the values of the surface diffusion coefficients for the combined diffusion model are about 15% smaller
than those for the effective diffusion coefficients. This difference is
attributed to the contribution of pore diffusion to the intraparticle
diffusion flux [5,15,16].

In multicomponent adsorption systems, because of interactive effects, displacement becomes a characteristic phenomenon, as shown
in Figure 7.2. This figure provides an example of the phenol(1)+2,5-
dichlorophenol(2)/XAD-4 adsorption system with input liquid concentrations, $C_{01} = 1.0$ mmol/l, $C_{02} = 1.0$ mmol/l; surface diffusion coefficients, for solid diffusion model, $D_{s1} = 3.0 \times 10^{-8}$ cm$^2$/sec,
$D_{s2} = 1.5 \times 10^{-8}$ cm$^2$/sec; for combined diffusion model, $D_{s1} =
2.6 \times 10^{-8}$ cm$^2$/sec, $D_{s2} = 1.2 \times 10^{-8}$ cm$^2$/sec.

In this binary system, phenol has a higher intraparticle diffusion
rate and will initially diffuse into the particle faster than 2,5-dichloro-
phenol; its adsorbed amount increases until a maximum is reached.
The accumulation in the solid phase then declines towards the equilibrium condition. That part of the curve that exceeds the final equilibrium point, "excess curve," represents the phenomenon of displace-

**Table 7.3.**   Statistical Results of Multicomponent Adsorption
Equilibrium

| System | Data No. | 3-P $S_1$ | $S_2$ | $S_3$ | IAS $S_1$ | $S_2$ | $S_3$ |
|---|---|---|---|---|---|---|---|
| Phenol(1)+ PCP(2)/ XAD-4 | 21 | 23.0 | 9.6 | | 12.3 | 6.2 | |
| Phenol(1)+ 2,5-DCP(3)/ XAD-4 | 20 | 10.3 | | 16.2 | 10.8 | | 15.0 |
| Phenol(1)+ PCP(2)+ 2,5-DCP(3)/ XAD-4 | 22 | 38.6 | 6.5 | 11.2 | 13.7 | 7.8 | 14.3 |

$S_i$ is a standard deviation defined as:

$$S_i = \sqrt{\frac{\sum_{j=1}^{n}(1 - (q_{i,cal}/q_{i,exp})_j)^2}{n}} \times 100\%$$

ment during the adsorption process. From Figure 7.2, it is noted that
only the combined diffusion model can describe this displacement effect. The solid diffusion model can not produce a displacement curve.

Figure 7.3 illustrates model-predicted concentration profiles along
particle radius. The combined diffusion model describes how the
weakly adsorbed component competes with and is replaced by the
strongly adsorbed component. For the solid diffusion model, no displacement phenomenon can be observed.

If the concentration profiles within the particle are integrated according to time, then kinetic curves are produced as shown in Figure
7.4. Although surface diffusion is a major rate controlling factor in
binary adsorption, pore diffusion plays an important role during the
competitive stage of adsorption, because the weakly adsorbed com-

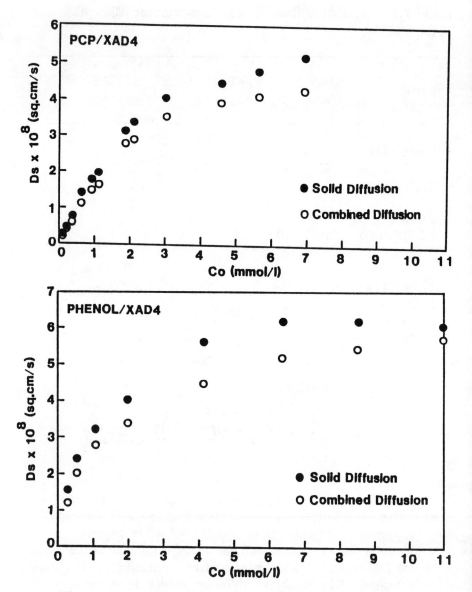

Figure 7.1.    $D_s$ vs. Concentration for Different Systems

ponent is more mobile after being displaced. Without the pore diffusion term, the model cannot describe the difference in the movement of the substituted molecules, and cannot produce displaced curves.

*Modeling Simulation*

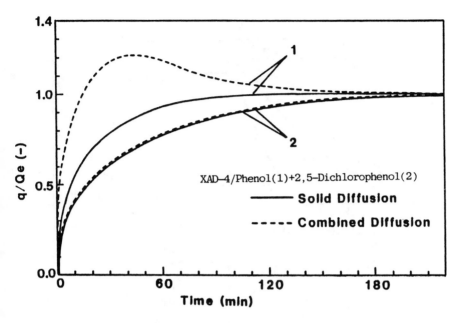

**Figure 7.2.**   Predicted Binary Kinetic Curves Using Solid Diffusion
and Combined Diffusion Models

To demonstrate the validity of the two models, model simulations
were compared to experimental data. Experimental studies were carried out for two binary systems: phenol+$p$-chlorophenol, phenol+2,5-
dichlorophenol, and one ternary system: phenol+$p$-chlorophenol+2,5-
dichlorophenol. Figures 7.5 and 7.6 represent the kinetic curves for
the phenol+PCP/XAD-4 adsorption system.

In this system, phenol which is weakly adsorbed , diffused faster at
the beginning; later it had to compete with $p$-chlorophenol for adsorption sites. Under these conditions, the adsorbed amount was reduced
by displacement by the strongly adsorbed $p$-chlorophenol. Simulated
kinetic curves show that the experimental kinetic data can best be
described by the combined diffusion model. Both models can produce
satisfying curves for $p$-chlorophenol. However, only the combined diffusion model agrees with the experimental data for phenol.

Because phenol and 2,5-dichlorophenol have very different adsorptive behaviors, two examples for this binary adsorption system are
illustrated in Figures 7.7 and 7.8.

In this system, phenol has a higher pore and surface diffusivity,
but a lower adsorption capacity than 2,5-dichlorophenol does. Model

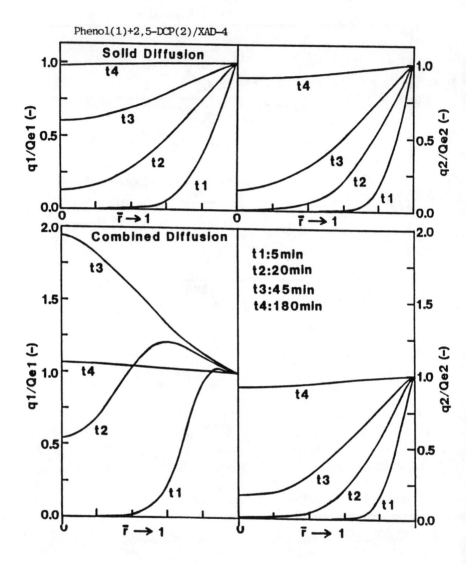

**Figure 7.3.**    Simulated Concentration Profiles along Particle
Radius at Different Times

simulations show that the combined diffusion model describes these
experimental kinetic data quite well.

The results also showed that the displacement effect was reduced
when the molar ratio between the weakly and strongly adsorbed components was low or high. When the molar ratio is low, the intraparticle

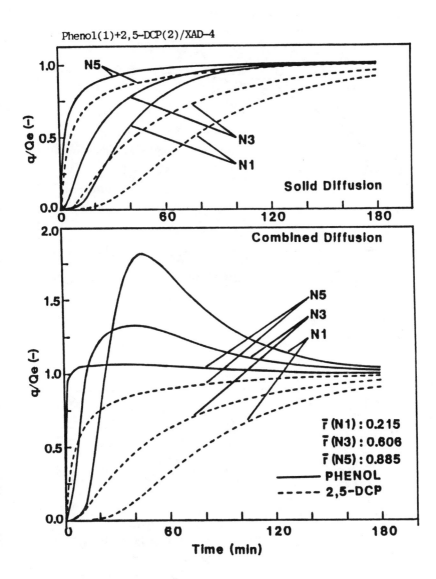

**Figure 7.4.** Simulated Concentration Profiles at Different Particle
Radial Positions

diffusion flux of the strongly adsorbed component may be larger than
that of the weakly adsorbed component; therefore, competition occurs from the beginning of the adsorption process and there will be
no visible excess curve. If the molar ratio is high, the weakly adsorbed component will not be affected significantly by the strongly

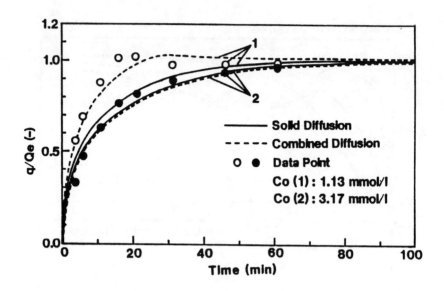

**Figure 7.5.**    Comparison of Model Simulations for Phenol(1)+ PCP(2)/XAD-4

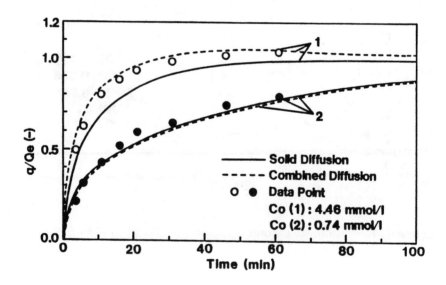

**Figure 7.6.**    Comparison of Model Simulations for Phenol(1)+ PCP(2)/XAD-4

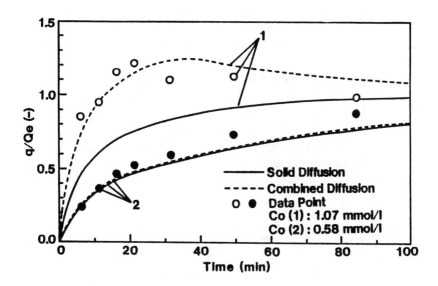

**Figure 7.7.**    Comparison of Model Simulations for Phenol(1)+
2,5-DCP(2)/XAD-4

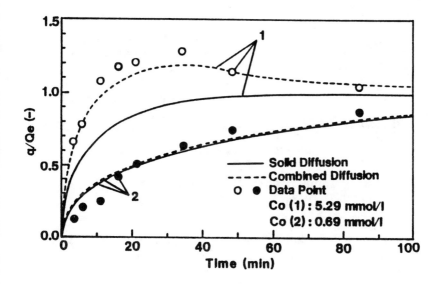

**Figure 7.8.**    Comparison of Model Simulations for Phenol(1)+
2,5-DCP(2)/XAD-4

adsorbed component during the adsorption process. For the usual range of molar ratios, (both molar amounts comparable), the excess curve is apparent and the excess amount increases when the molar ratio increases.

For an adsorption system in which more than two components co-exist, the procedure of estimating diffusion parameters becomes more complicated. Since the combined diffusion model can describe the competitive adsorption process, the model can be extended to multicomponent adsorption systems easily. Figures 7.9 and 7.10 compare the kinetic curves generated by experiment and by modeling simulation, for the adsorption system of phenol+$p$-chlorophenol+2,5-dichlorophenol/XAD-4.

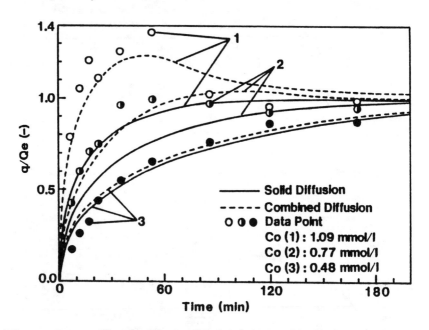

**Figure 7.9.**  Comparison of Model Simulations for Phenol(1)+ PCP(2)+2,5-DCP(3)/XAD-4

Obviously, a more complicated situation can be expected in the particle during the adsorption process. In this system, phenol is the most weakly adsobed component, although its diffusion rate is faster than the other components. However, the adsorbed molecules are easily displaced by the other components. $p$-Chlorophenol has adsorptive characteristics between the other two adsorbates; therefore, it displaces phenol and then is displaced by 2,5-dichlorophenol until the final equilibrium condition is reached. The most strongly adsorbed

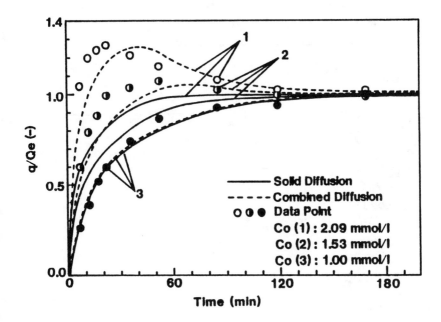

**Figure 7.10.**    Comparison of Model Simulations for Phenol(1)+
PCP(2)+2,5-DCP(3)/XAD-4

component, 2,5-dichlorophenol, diffuses slowly, but its accumulation
in the solid phase gradually increases until equilibrium is reached. As
a result of this displacement sequence, the excess curve of phenol appears first and is more significant than that of $p$-chlorophenol. There
is no excess curve for 2,5-dichlorophenol.

Table 7.4 lists the statistical results of modeling simulation for the
three multicomponent adsorption systems in this study. It can be seen
that the combined diffusion model usually performs better, and the
percent standard deviations are comparable to those for the equilibrium studies for the the same adsorption system. Because the solid
diffusion model fails to account for competition and displacement effects, the modeling simulation is poor.

## CONCLUSIONS

Modeling simulation indicates that the combined diffusion model
describes the displacement and competition phenomena in the multicomponent adsorption systems in a reasonable manner. The solid
diffusion model, which is widely applied in liquid-solid adsorption sys-

# ADSORPTION TECHNOLOGY

**Table 7.4.** Statistical Results of Multicomponent Adsorption Kinetics

| System | Data No. | SD | | | CD | | |
|---|---|---|---|---|---|---|---|
| | | $S_1$ | $S_2$ | $S_3$ | $S_1$ | $S_2$ | $S_3$ |
| Phenol(1)+ PCP(2)/ XAD-4 | 175 | 11.6 | 3.9 | | 4.6 | 4.2 | |
| Phenol(1)+ 2,5-DCP(3)/ XAD-4 | 66 | 31.7 | | 7.1 | 8.7 | | 6.8 |
| Phenol(1)+ PCP(2)+ 2,5-DCP(3)/ XAD-4 | 88 | 34.9 | 18.2 | 4.9 | 11.5 | 8.9 | 5.2 |

* Molecular diffusivities in the combined diffusion model were estimated according to the Wilke-Chang correlation [17]: $D_{m1} = 1.01 \times 10^{-5}$ cm$^2$/sec, $D_{m2} = 9.2 \times 10^{-6}$ cm$^2$/sec, $D_{m3} = 7.6 \times 10^{-6}$ cm$^2$/sec.
SD = Solid Diffusion Model; CD = Combined Diffusion Model.

tems, does not simulate these displacement phenomena. From a theoretical point of view, the solid diffusion model is applicable to diffusion in homogeneous material or where the adsorbates in a mixture have similar adsorptive behaviors. Adsorption in a porous adsorbent, the heterogeneous case, is better described by the combined diffusion model.

The excess curve is a phenomenon that resulted from adsorption displacement and competition effects. When the molar ratio between the weakly and strongly adsorbed components increases, the excess amount increases and the time of appearance is delayed. The displacement phenomenon is more significant when the adsorbates in a mixture have different adsorptive characteristics.

*REFERENCES*

1. Aguwa, A.A. *Evaluation of Sorption Behavior of Single and Multicomponent Organics with Synthetic Resin.* Ph.D. Thesis, Illinois Institute of Technology, Chicago, IL (1984).
2. Ferret, G.J. *Sorption Behavior of Phenol and 2,5-Dichlorophenol on Synthetic Resin.* M.S. Thesis, Illinois Institute of Technology, Chicago, IL (1984).
3. Dinopoulou, G. *Sorptive Behavior of a Multicomponent Solution of Phenolics with Synthetic Resin.* M.S. Thesis, Illinois Institute of Technology, Chicago, IL (1985).
4. Mathews, A.P., and C.A. Su, *Envir. Prog.*, **2**, 257 (1983).
5. Fritz, W., W. Merk, and E.U. Schlünder, *Chem. Eng. Sci.*, **36**, 731 (1981).
6. Liapis, A.I., and D.W.T. Rippin, *Chem. Eng. Sci.*, **32**, 619 (1977).
7. Merk, W., W. Fritz, and E.U. Schlünder, *Chem. Eng. Sci.*, **36**, 743 (1981).
8. Palekar, M.G., and R.A. Rajadhyaksha, *Chem. Eng. Sci.*, **41**, 463 (1986).
9. Villadsen, J.V. and W.E. Stewart, *Chem. Eng. Sci.*, **22**, 1483 (1967).
10. Finlayson, B.A. *The Method of Weighted Residuals and Variational Principles.* Academic Press, New York, NY (1972).
11. Crittenden, J.C., B.W.C. Wong, W.E. Thacher, V.L. Snoeyink, and R.L. Hinrichs, *J. WPCF.*, **52**, 2780 (1980).
12. Mathews, A.P. *Mathematic Modeling of Multicomponent Adsorption in Batch Reactors.* Ph.D. Thesis, University of Michigan, Ann Arbor, MI (1975).
13. Dullien, F.A.L. *AIChE. J.*, **21**, 299 (1975).
14. Dullien, F.A.L. *AIChE. J.*, **21**, 820 (1975).
15. Komiyama, H., and J.M. Smith, *AIChE. J.*, **20**, 728 (1974).
16. Neretnieks, I. *Chem. Eng. Sci.*, **31**, 1029 (1976).
17. Wilke, C.R., and P. Chang, *AIChE. J.*, **1**, 264 (1955).

## CHAPTER VIII

---

## FIXED-BED MODELING AND DESIGN
## FOR WATER ADSORPTION SYSTEMS

---

### INTRODUCTION

The application of the adsorption process for water treatment is usually through application of a fixed bed. Because of unsteady-state conditions during the adsorption process, a mathematical model is required for design. Equilibrium and kinetic data derived under simplified conditions can provide fundamental information needed to verify the models. Therefore, a systematic study of fixed-bed adsorbers is presented in this chapter.

A widely applied model is the solid diffusion model [1-3]. However, because the adsorption process takes place in a porous particle, a pore diffusion model was developed to describe the diffusion in the pore network of the particle [4-6]. To interpret the migration of molecules through the pore space and pore walls [7,8], the combined diffusion model includes both pore diffusion and surface diffusion mechanisms [9-11]. Because the effects of incomplete equilibrium and nonlinear characteristics need to be taken into account, the modeling equations are solved numerically. Although the implicit finite difference scheme, the Crank-Nicholson method, has been widely applied to solve the equations in the models [12-14], the orthogonal collocation method [15,16] has been shown to be better for solving fixed bed models [17-19]. Therefore, the orthogonal collocation method was applied for numerical analysis.

### EXPERIMENTAL RESULTS AND DISCUSSION

#### Mathematical Model

The combined diffusion model for the material balance in the particle, Equation (2.50) was applied for the cases of favorable isotherm and dilute solution range. Since the Peclet numbers show $Pe(= VL/D_{ax}) > 40$ and $Pe_p(= 2R_pV/D_{ax}) \to 2.0$ in this study, longitudinal dispersion is negligible [20,21] in the material balance for the fluid phase. In order to apply a numerical technique, the partial differential equations were transfered to the following dimensionless equations:

Particle:

$$\frac{\partial Y}{\partial T} = \frac{N_s}{\bar{r}^2} \frac{\partial}{\partial \bar{r}} \left( \bar{r}^2 \frac{\partial Y}{\partial \bar{r}} \right) + \alpha_p \frac{N_p}{\bar{r}^2} \frac{\partial}{\partial \bar{r}} \left( \bar{r}^2 \frac{\partial X_p}{\partial \bar{r}} \right) \qquad (8.1)$$

Fixed bed:

$$\frac{\partial X}{\partial T} = -\frac{1}{\alpha_b} \frac{\partial X}{\partial \bar{z}} - \frac{1}{\alpha_b} \frac{\partial Y_{av}}{\partial T} \qquad (8.2)$$

Initial and boundary conditions:

$$St(X - X_s) = N_s \frac{\partial Y}{\partial \bar{r}} + \alpha_p N_p \frac{\partial X_p}{\partial \bar{r}} \qquad \text{at } \bar{r} = 1 \qquad (8.3)$$

$$St(X - X_s) = \frac{\partial}{\partial T} \int_0^1 Y \bar{r}^2 \, d\bar{r} \qquad (8.4)$$

$$\partial Y / \partial \bar{r} = 0, \quad \partial X_p / \partial \bar{r} = 0 \qquad \text{at } \bar{r} = 0 \qquad (8.5)$$

$$Y = 0, \quad X_p = 0, \quad X = 0 \qquad \text{when } T = 0 \qquad (8.6)$$

$$X = 1, \quad \partial X / \partial T = 0 \qquad \text{at } \bar{z} = 0, \text{ when } T > 0 \qquad (8.7)$$

where

$$Y = q/Q_e; \qquad Y_{av} = q_{av}/Q_e$$

$$X = C/C_0; \qquad X_p = C_p/C_0; \qquad X_s = C_s/C_0$$

$$\bar{r} = r/R_p; \qquad T = t\alpha_b V/L; \qquad \bar{z} = z/L$$

$$\alpha_b = C_0 \epsilon_b / Q_e \rho_p (1 - \epsilon_b); \qquad \alpha_p = C_0 \epsilon_p / Q_e \rho_p$$

$$N_s = D_s L / R_p^2 \alpha_b V; \qquad N_p = D_p L / R_p^2 \alpha_b V$$

$$Y_{av} = 3 \int_0^1 Y \bar{r}^2 \, d\bar{r}; \qquad St = K_f L (1 - \epsilon_b) / R_p \epsilon_b V$$

Equation (8.3) states that the diffusion flux for the external liquid film is equal to the intraparticle diffusion flux. Equation (8.4) states that the mass transfer flux at the interphase between the liquid and the particle is equal to the time rate of change of the average accumulation of sorbate inside the particle. Equations (8.5) result from symmetry. The local equilibrium adjacent to the exterior surface of the particle and the interior pore walls can be described by the Freundlich type equation:

$$q_e = K C_e^{1/n} \qquad (8.8)$$

Its dimensionless form is:

$$Y_e = X_e^{1/n} \tag{8.9}$$

By applying the orthogonal collocation method, the second order partial differential equation describing instantaneous mass balance inside particles can be computed approximately by the following set of first order ordinary differential equations:

$$\left(\frac{dY}{dT}\right)_{jk} = N_s\left(\sum_{n=1}^{N+1} B_{jn}Y_{nk}\right) + \alpha_p N_p\left(\sum_{n=1}^{N+1} B_{jn}X_{pnk}\right) \tag{8.10}$$

After substituting Equation (8.4) and integrating form a of $Y_{av}$ into Equation (8.2), the first order partial differential equation for mass balance in a fixed bed can be approached by a set of first order ordinary differential equations:

$$\left(\frac{dX}{dT}\right)_k = \frac{-1}{\alpha_b}\left(\sum_{m=1}^{M+2} A_{km}X_m\right) - 3\frac{St}{\alpha_b}(X_k - X_{sk}) \tag{8.11}$$

The mass transfer at particle surface from Equation (8.4) can be calculated by:

$$St(X_k - X_{sk}) = \sum_{n=1}^{N+1} W_n\left(\frac{dY}{dT}\right)_{nk} \tag{8.12}$$

where $A$ is an $(M + 2, M + 2)$ matrix of collocation coefficients for cylindrical geometry system; $B$ is an $(N + 1, N + 1)$ matrix of collocation coefficients for the spherical geometry system; $W$ is a set of weighting coefficients; $j$ and $n$ are the collocation numbers within the particle; $k$ and $m$ are the collocation numbers along the column. The program written in Fortran IV language was solved by the GEAR package that was developed by Hindmarsh [22].

*Experimental Procedure*

The two adsorbates used in this study are phenol and p-chlorophenol. The adsorbents are Amberlite resin XAD-4 and Calgon's activated carbon F400. The properties of the adsorbents and adsorbates have been discussed in Chapter V.

Ultraviolet Spectrophotometer (Perkin-Elmer) and 10-mm pathlength rectangular glass cells with standard silica windows (Fisher

Scientific Company) were used in the analyses of the liquid concentration. The optimum wavelength used was 269.5 nm for phenol and 279.7 nm for p-chlorophenol aqueous solutions.

For the fixed bed experiments, a plexiglass column of 2.46 cm ID and 30.0 cm length was used. After immersing a segment of column in phenolic solutions for 24 hours, it was found that the column material did not adsorb either of the adsorbates. Five sampling positions along the column were at 6 cm, 12 cm, 18 cm, 24 cm and 30 cm from the bottom (see Figure 8.1).

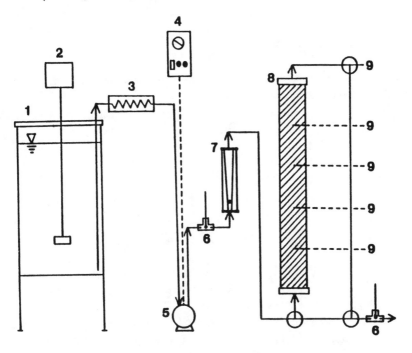

1. RESERVOIR TANK        6. THERMOMETER
2. STIRRING MOTOR        7. FLOWMETER
3. HEAT EXCHANGER        8. COLUMN
4. SPEED CONTROLLER      9. SAMPLING POSITION
5. FEED PUMP

Figure 8.1.    Column Adsorption Experimental System

Capillary sampling tubes with 1.0 mm OD and 0.35 mm ID reached the center of the column at each sampling position. Adsorbent particles were packed into the column and allowed to settle for 24 hours to make sure that there were no bubbles in the column. Fiberglass was packed on the top of the column to fix the adsorbent particles. The ratio of column radius to particle radius is larger than 21 so that the plug flow profile was assumed [23]; entrance and wall effects were minimized under this condition [24]. The hydraulic loading regime was 2.5–10.0 GPM/ft$^2$, which was believed to be the normally accepted loading regime [25]. Solution was input into the column by using a positive-displacement pump of Teflon, Model SP40002 (Fluorocarbon Company) which was controlled by an adjustable speed control, Model SL-32 (Minarik Electric Company). All experiments were conducted at room temperature, $24 \pm 2°$ C.

*Adsorption Equilibrium and Kinetic Parameters*

The adsorption equilibrium and kinetic data were determined by the gravimetric method (Chapter V). The effective surface diffusion coefficients for the surface diffusion model were estimated and shown in Chapter V and those for the combined diffusion model were discribed in Chapter VI. For the combined diffusion model, the relative contribution of surface diffusion to intraparticle mass transfer was estimated as follows:

$$R_s = \frac{D_s}{D_s + \dfrac{\epsilon_p D_p}{\rho_p}\left(\dfrac{\partial C_p}{\partial q}\right)_{mean}} \tag{8.13}$$

This ratio was found to be about 80% for the phenol/XAD-4 and $p$-chlorophenol/XAD-4 systems, 95% for the phenol/F400 and $p$-chlorophenol/F400 systems. The reasons for these different ratios result from the properties of the adsorbents; XAD-4 has larger pores and a weaker adsorptive behavior than F400 does. Because surface diffusion represents a major rate-controlling factor for intraparticle mass transfer, it is reasonable to use the solid diffusion model to describe some solid-liquid adsorption systems.

*Sensitivity Analysis*

Operational factors such as input liquid concentration, particle size, and flow rate are important in long column designs. In order to determine which parameter might have a significant effect on the breakthrough curves, a sensitivity analysis using the combined diffusion model was carried out and provided the following results:

1. Collocation numbers: The accuracy of the orthogonal collocation method depends on the selected collocation numbers. For the fixed-

bed problem, the testing procedure involves two steps: one for collocation number $M$ along the column longitudinal coordinate and one for $N$ along the particle radial coordinate. An example is illustrated in Figure 8.2.

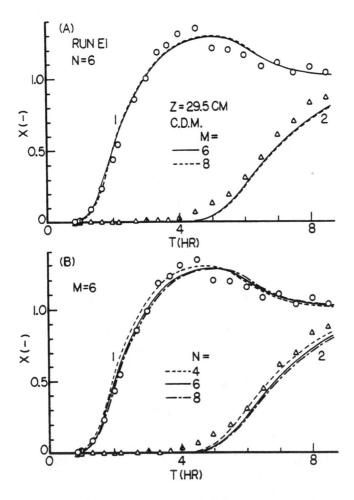

Figure 8.2.    The Effect of Collocation Number

Results show that the deviation between $M = 6$ and 8 is much smaller than that between $M = 4$ and 6, and the difference between $N = 6$ and 8 is less significant than that between $N = 4$ and 6; therefore, $M = 6$ and $N = 6$ are sufficient.

2. Isotherm parameters: The effect of variation in equilibrium parameters was tested through modeling simulations, as shown in Figure

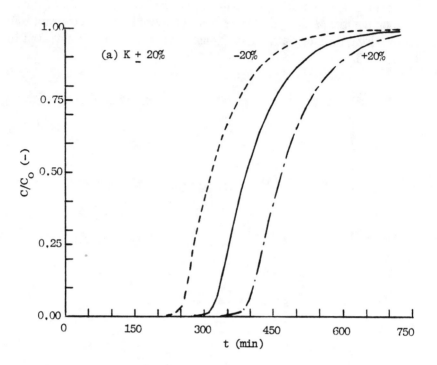

**Figure 8.3.**    The Effect of Isotherm Parameters

8.3. Results show that the predicted breakthrough curve shifts along the time scale proportional to the calculated equilibrium capacity. This effect is more significant when the value of the solute partition coefficient $P_s$ is larger.

$$P_s = \frac{Q_e \rho_p}{C_0 \epsilon_p} \qquad (8.14)$$

Therefore, the isotherm determines the breakthrough time.

3. Intraparticle diffusion coefficients: Since surface diffusivity is a major rate-controlling factor, the effect of errors in this parameter are more apparent than for pore diffusivity, as shown in Figure 8.4. A larger value of diffusivity produces a sharper breakthrough curve because larger intraparticle diffusion flux can reduce the length of the mass transfer zone while the moving velocity of the adsorption zone remains the same.

4. External mass transfer coefficient: Results show that deviations in this parameter have a minor effect, as shown in Figure 8.5. A larger value for the film mass transfer coefficient will generate a

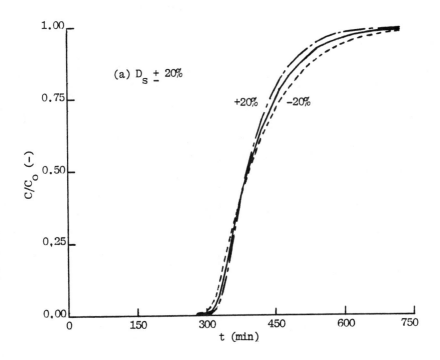

**Figure 8.4.**    The Effect of Diffusivities

sharper breakthrough curve. To judge the relative effect between the external and internal diffusion resistances, a Biot number $Bi$ was defined as follows [26].

$$Bi = \frac{C_0 k_f R_p}{D_p C_0 + D_s \rho_p Q_e} \tag{8.15}$$

Results showed that the effect of $k_f$ is more important when the Biot number is smaller; therefore, the impact of flow rate is more significant under this condition.

5. Particle size: Since spherical and uniform particles are usually assumed in modeling simulation, the sphericity and size distribution of particles can affect the estimation of effective particle size [27,28]. It has been found that the breakthrough curve generated from narrow and normal size distributions of particles is very close to that from uniform-sized particles [29]. Since smaller particles can reduce film mass transfer resistance and the intraparticle diffusion path, the length of the mass transfer zone is reduced and a sharper breakthrough curve can be observed for smaller particles.

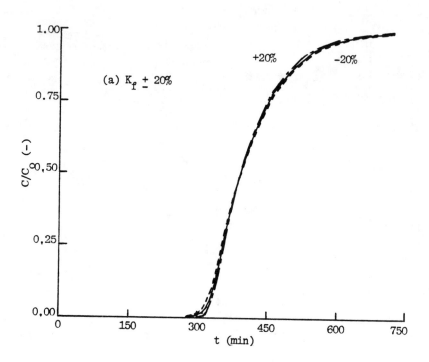

**Figure 8.5.** The Effect of External Mass Transfer Coefficient and Particle Size

*Modeling Comparison*

The simplest model to predict breakthrough time is the equilibrium model. If the effect of mass transfer resistances is neglected, then a vertical wave front is formed and its breakthrough time at column length $L$ is determined directly by column capacity, as follows:

$$t_w = \frac{W_s L Q_e}{V \epsilon_b C_0} \tag{8.16}$$

where $W_s$ is the dosage of adsorbent per unit volume of bed.

Examples of applying this model are shown in Figure 8.6. Experimental and analytical parameters are listed in Tables 8.1–8.3, respectively.

Note that in these adsorption systems, although the equilibrium model could predict approximately the appearance time of the breakthrough curves, it could not generate a pattern similar to the experimental data. This deviation shows that the mass transfer resistance

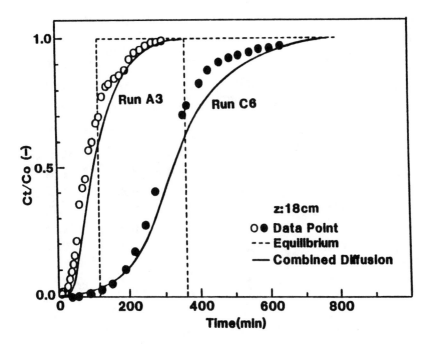

**Figure 8.6.**   Comparison between the Equilibrium and Combined
Diffusion Model

is important in the design of an adsorption system. The length of
the adsorption zone is longer when the mass transfer resistances are
larger, because a longer time is needed to reach equilibrium.

Figures 8.7 and 8.8 illustrates the comparison of predicted and experimental breakthrough curves. Both solid diffusion and combined
diffusion models can generate curves which have the same pattern as
the real effluent concentration profile, because of the effects of external and intraparticle diffusion parameters in the models. Nevertheless,
there are some differences between the two modeling simulations. If
an adsorption system has a large contribution of pore diffusion in the
total intraparticle diffusion, or its Biot number is so large that the
simulation is sensitive to the intraparticle diffusion parameter, then
the deviation between the models will be significant.

## BREAKTHROUGH CURVES

To demonstrate the applicability of the model under different operative conditions, fixed bed experiments were conducted using different
input concentrations, flow rates and bed depths, as shown in Table

**Table 8.1.** Experimental Parameters of Fixed Bed Adsorption – Single Component System

| Run* | $C_0$ (mg/l) | FLRT (GPM/ft$^2$) | Temp ($^0$C) | $\epsilon_b$ (–) | $V$ (cm/sec) |
|------|------|------|------|------|------|
| A1 | 100.0 | 5.16 | 24.0 | 0.347 | 1.01 |
| A2 | 96.0 | 5.16 | 22.0 | 0.341 | 1.04 |
| A3 | 51.0 | 5.68 | 23.5 | 0.288 | 1.35 |
| A4 | 101.0 | 8.78 | 23.5 | 0.347 | 1.72 |
| A5 | 510.0 | 4.91 | 26.5 | 0.356 | 0.94 |
| A6 | 104.0 | 3.87 | 22.5 | 0.353 | 0.75 |
| B1 | 100.3 | 5.42 | 24.5 | 0.372 | 0.99 |
| B2 | 96.0 | 10.33 | 25.5 | 0.326 | 2.16 |
| B3 | 96.0 | 5.68 | 24.0 | 0.349 | 1.11 |
| B4 | 438.4 | 5.42 | 22.5 | 0.338 | 1.09 |
| B5 | 295.0 | 5.16 | 23.0 | 0.346 | 1.02 |
| B6 | 96.0 | 6.71 | 24.5 | 0.329 | 1.39 |
| C1 | 151.0 | 7.13 | 25.0 | 0.457 | 1.06 |
| C2 | 98.6 | 6.25 | 25.0 | 0.468 | 0.91 |
| C3 | 115.0 | 9.55 | 24.0 | 0.439 | 1.48 |
| C4 | 47.5 | 7.02 | 23.0 | 0.416 | 1.15 |
| C5 | 184.5 | 6.81 | 25.5 | 0.445 | 1.05 |
| C6 | 158.2 | 6.76 | 26.0 | 0.436 | 1.06 |
| D1 | 266.7 | 5.94 | 23.5 | 0.428 | 0.94 |
| D2 | 122.5 | 7.13 | 25.0 | 0.412 | 1.18 |
| D3 | 60.5 | 5.99 | 24.0 | 0.437 | 0.93 |
| D4 | 127.7 | 5.99 | 25.0 | 0.439 | 0.93 |
| D5 | 147.3 | 5.68 | 22.0 | 0.457 | 0.84 |
| D6 | 289.0 | 5.93 | 23.0 | 0.440 | 0.92 |

* System: A1 – A6 are Phenol/XAD4; B1 – B6 are $p$-Chlorophenol/ XAD4; C1 – C6 are Phenol/F400; D1 – D6 are $p$-Chlorophenol/ F400.

Table 8.2.   Analytic Parameters of Fixed Bed Adsorption –
Single Component System

| Run | $k_f$ $\times 10^3$ | $D_s$ $\times 10^8$ | $D_p$ $\times 10^6$ | $P_s$ | $Bi$ | $V_{mtz}$ $\times 10^3$ |
|---|---|---|---|---|---|---|
| A1 | 5.58 | 2.27 | 1.61 | 349 | 22.2 | 3.01 |
|    |      | 2.74 | (for Solid Diffusion Model) | | | |
| A2 | 5.33 | 2.23 | 1.52 | 355 | 21.6 | 2.95 |
| A3 | 4.98 | 1.74 | 1.59 | 463 | 19.7 | 2.31 |
| A4 | 6.46 | 2.28 | 1.59 | 348 | 25.8 | 5.16 |
| A5 | 5.68 | 4.32 | 1.68 | 177 | 22.9 | 5.72 |
| A6 | 4.88 | 2.30 | 1.49 | 344 | 19.9 | 2.32 |
| B1 | 5.68 | 1.24 | 1.44 | 826 | 19.2 | 1.36 |
|    |      | 1.56 | (for Solid Diffusion Model) | | | |
| B2 | 6.33 | 1.20 | 1.46 | 850 | 21.4 | 2.41 |
| B3 | 5.33 | 1.20 | 1.43 | 850 | 18.1 | 1.37 |
| B4 | 5.04 | 3.55 | 1.33 | 321 | 15.9 | 3.39 |
| B5 | 5.06 | 2.67 | 1.37 | 414 | 16.2 | 2.55 |
| B6 | 5.08 | 1.20 | 1.44 | 850 | 17.2 | 1.57 |
| C1 | 4.30 | 3.6 | 1.53 | 1741 | 5.3 | 0.826 |
|    |      | 3.8 | (for Solid Diffusion Model) | | | |
| C2 | 4.34 | 2.8 | 1.52 | 2466 | 4.9 | 0.523 |
| C3 | 4.61 | 3.2 | 1.52 | 2175 | 5.2 | 0.858 |
| C4 | 4.47 | 1.6 | 1.47 | 4478 | 4.9 | 0.295 |
| C5 | 4.68 | 4.5 | 1.52 | 1478 | 5.5 | 0.918 |
| C6 | 4.45 | 4.0 | 1.52 | 1676 | 5.2 | 0.788 |
| D1 | 3.60 | 5.7 | 1.34 | 1598 | 3.1 | 0.709 |
|    |      | 5.9 | (for Solid Diffusion Model) | | | |
| D2 | 3.76 | 3.3 | 1.36 | 3053 | 2.9 | 0.437 |
| D3 | 3.68 | 2.0 | 1.36 | 5492 | 2.7 | 0.212 |
| D4 | 3.70 | 3.4 | 1.36 | 2949 | 2.9 | 0.398 |
| D5 | 5.63 | 3.8 | 1.36 | 2619 | 2.3 | 0.435 |
| D6 | 3.62 | 6.3 | 1.31 | 1495 | 3.1 | 0.779 |

**Table 8.3.**   Physical Parameters of Fixed Bed Adsorption –
Single Component System

| Run | $Re$ (–) | $Sc$ (–) | $D_{ax}^{\dagger}$ (cm$^2$/sec) | $Pe/L$ (cm$^{-1}$) | $Pe_p$ (–) | $R_{col}/R_p$ (–) |
|-----|------|------|-------|------|------|----|
| A1 | 4.78 | 971 | 0.023 | 44.3 | 1.95 | 55 |
| A2 | 4.68 | 1083 | 0.024 | 43.4 | 1.91 | 55 |
| A3 | 5.80 | 993 | 0.029 | 45.3 | 2.0 | 55 |
| A4 | 7.56 | 993 | 0.038 | 45.5 | 2.0 | 55 |
| A5 | 4.86 | 903 | 0.022 | 41.1 | 1.81 | 55 |
| A6 | 3.11 | 1128 | 0.017 | 44.4 | 1.95 | 55 |
| B1 | 4.46 | 1082 | 0.022 | 45.5 | 2.0 | 55 |
| B2 | 9.96 | 1056 | 0.048 | 45.2 | 1.99 | 55 |
| B3 | 5.02 | 1084 | 0.024 | 44.0 | 1.94 | 55 |
| B4 | 4.73 | 1260 | 0.025 | 44.0 | 1.94 | 55 |
| B5 | 4.60 | 1205 | 0.023 | 44.7 | 1.97 | 55 |
| B6 | 5.57 | 1199 | 0.031 | 43.5 | 1.91 | 55 |
| C1 | 11.4 | 967 | 0.054 | 19.5 | 1.93 | 24 |
| C2 | 9.82 | 969 | 0.047 | 19.4 | 1.95 | 24 |
| C3 | 15.8 | 971 | 0.075 | 19.7 | 1.97 | 24 |
| C4 | 11.9 | 1036 | 0.059 | 19.5 | 1.94 | 24 |
| C5 | 11.4 | 967 | 0.055 | 19.1 | 1.99 | 24 |
| C6 | 11.1 | 967 | 0.053 | 19.8 | 1.99 | 24 |
| D1 | 10.1 | 1109 | 0.049 | 19.2 | 1.96 | 24 |
| D2 | 12.8 | 1080 | 0.061 | 20.0 | 1.96 | 24 |
| D3 | 10.1 | 1084 | 0.048 | 19.8 | 1.98 | 24 |
| D4 | 10.1 | 1080 | 0.048 | 19.7 | 1.97 | 24 |
| D5 | 4.29 | 1262 | 0.022 | 37.3 | 1.95 | 49 |
| D6 | 9.65 | 1157 | 0.048 | 19.3 | 1.94 | 24 |

† Values estimated by the following equation [30]:

$$D_{ax} = \frac{D_m}{\epsilon_b}(20.0 + 0.5 ReSc)$$

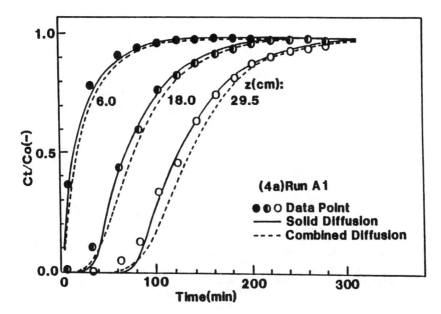

Figure 8.7.    Comparison between the Surface Diffusion and
Combined Diffusion Model- Example A

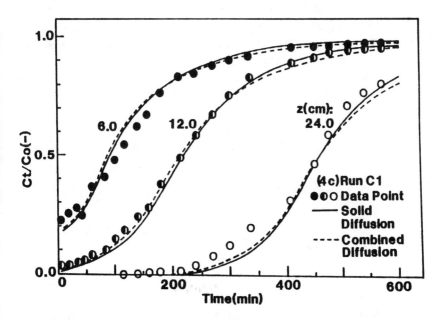

Figure 8.8.    Comparison between the Surface Diffusion and
Combined Diffusion Model- Example B

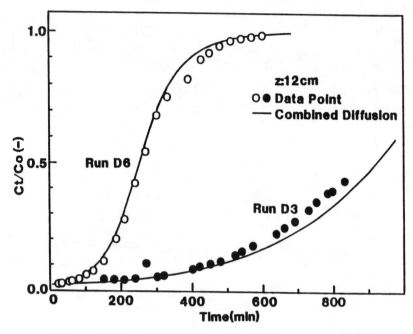

Figure 8.9.    The Effect of Input Concentration

8.1. Since solute partition coefficient, $P_s$, is an index of separation, and Biot number, $Bi$, indicates which mass transfer parameter is a rate-controlling factor, they provide useful information on fixed-bed performance. Another parameter used in this analysis is the moving velocity of the mass transfer zone, $V_{mtz}$, which was derived from equilibrium assumptions, as follows [31]:

$$V_{mtz} = \left(\frac{\partial z}{\partial t}\right)_C = \frac{V}{1 + \rho_p \dfrac{1 - \epsilon_b}{\epsilon_b}\left(\dfrac{\partial \overline{q}}{\partial C}\right)_C} \qquad (8.17)$$

The value of $\partial \overline{q}/\partial C$ can be estimated as $Q_e/C_0$. Results of those analytical parameters are listed in Table 8.2. In this table, $k_f$ values were estimated by using Equation (2.31), units of parameters are $k_f$: [cm/sec], $D_s$: [cm$^2$/sec], $D_p$: [cm$^2$/sec], $P_s$: [- ], $Bi$: [- ], $V_{mtz}$: [cm/sec].

Under practical operating conditions, the adsorption process may encounter different concentrations of solution. Therefore, the effects of variation in input concentration were studied for different adsorption systems, some examples of which are shown in Figure 8.9.

Note that the breaktime for higher input concentration is always earlier than that for lower input concentration. The reason for this phenomenon can be explained through Equation (8.17). For a linear isotherm adsorption system, the value of $\partial \bar{q}/\partial C$ is a constant, so that the moving velocity of the MTZ is a constant and the breaktime will not be affected by different input concentrations. However, the adsorption systems in this study have a nonlinear, favorable isotherm. As the input concentration increases, the value of $\partial \bar{q}/\partial C$ decreases so that the zone velocity increases; therefore, the breaktime is shorter under this circumstance.

From Figure 8.9, it is also found that the shape of the breakthrough curve is steeper at higher input concentration than that at lower input concentration. This is evidence that the intraparticle diffusivity varies with liquid concentration, because modeling analyses have shown that the mass transfer resistances affect the shape of the breakthrough curve. This is in good agreement with the results derived from differential reactor kinetic curves (Chapter V). Therefore, a sharper breakthrough curve generated at higher input concentration essentially results from a larger intraparticle diffusivity, because the adsorption zone is reduced by higher diffusion flux.

Equation (9.17) also indicates that the moving velocity of the MTZ is a function of interstitial velocity. Since the flow rate is an important factor in fixed bed design, its effects were observed by testing different fluid velocities, as illustrated in Figure 8.10. Results show that the breaktime was shorter when a higher flow rate was employed.

Usually, the intraparticle diffusivity is believed to be independent of flow rate [32]. From Figure 8.10, however, the breakthrough curves were found to be steeper at higher flow rate. This phenomenon is attributed to external film mass transfer resistance. This resistance is smaller when flow rate is higher, so that the length of mass transfer zone is reduced and a sharper breakthrough curve is generated.

If the breakthrough curves are obtained at different column lengths, their shape should be similar because all of the other operative conditions are the same. However, the experimental results show that the breakthrough curve has a sharper shape for a shorter column. If the column length is shorter than the zone length, the liquid concentration can break through the column immediately. An abrupt and incomplete breakthrough curve is formed, as shown in Figures 8.7 and 8.8. The concentration profiles in the MTZ will be stable after travelling some finite distance, which is usually much longer than the zone length [23]. Therefore, a constant pattern of breakthrough curve can occur only when the column is of sufficient length [14].

If different adsorption systems having the same input concentration and interstitial velocity are compared, as shown in Figure 8.11,

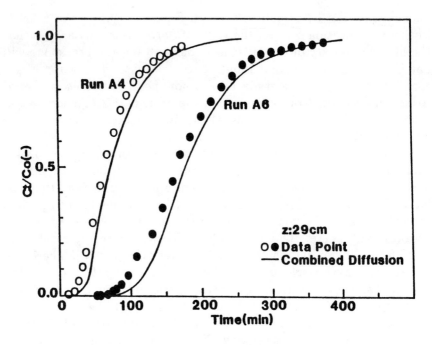

Figure 8.10.　The Effect of Flow Rate

the breakthrough curves from resin beds have a sharper shape and appear earlier than those from carbon beds. Because the resin adsorbers usually have a smaller value of solute partition coefficient, a larger Biot number and faster moving velocity for the MTZ than the carbon adsorbers do, a smaller solute partition coefficient results from less affinity between adsorbate and adsorbent. Therefore, column capacity is lower and the process has less efficiency of separation. A larger Biot number means smaller solute partition in the solid phase and a larger relative value between the external and intraparticle mass transfers. Consequently, the adsorbent reaches saturation more easily and a sharper shape for the breakthrough curve is obtained. A faster moving velocity for the MTZ is due to smaller solute distribution in the solid phase. Therefore, the breakthrough curve appears earlier at a specified column length.

*Empirical Method*

Because the numerical procedure for calculation of the breakthrough time is complicated, a quick estimation of the breaktime can be derived from experimental data. According to modeling analysis,

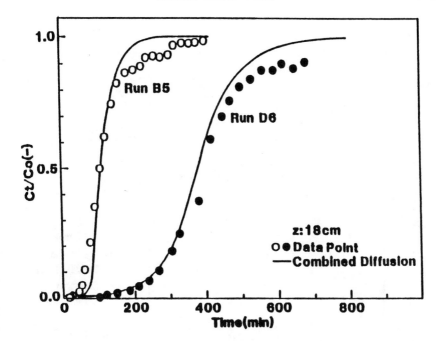

**Figure 8.11.**  Comparison between Different Adsorption Systems

the slope of the breakthrough curve at $X = 0.5$ was correlated with intraparticle diffusion resistance, $\overline{D}_{int}/R_p^2$, external diffusion resistance, $K_f a_p$, solute partition in bed, $P_b$, and bed void parameter, $\epsilon_b/(1 - \epsilon_b)$, as follows.

$$\left(\frac{dX}{dt}\right)_{X=0.5} = \left(\frac{\overline{D}_{int}k_f a_p \epsilon_b}{R_p^2 P_b(1 - \epsilon_b)}\right)^{1/2} \qquad (8.18)$$

where

$$a_p = 3(1 - \epsilon_b)/R_p \qquad (8.19)$$

$$P_b = 1/\alpha_b \qquad (8.20)$$

$$\overline{D}_{int} = D_s + D_p \alpha_p \qquad (8.21)$$

Then the breaktime at $z = L$ and $X = 0.1$ can be estimated as follows.

$$t_b = \frac{L}{V_{mtz}} - \frac{0.4}{(dX/dt)_{X=0.5}} \qquad (8.22)$$

Some examples showing the application of this method are illustrated in Figure 8.12.

A comparison of observed and calculated data is shown in Figure 8.13 and Table 8.4. The standard deviation of the results was found

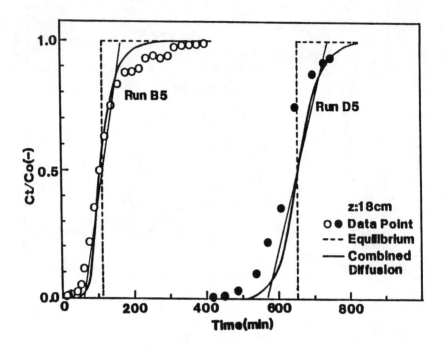

**Figure 8.12.**    Application of the Empirical Method

to be smaller for longer columns and in carbon beds, because the relative error was smaller when the breaktime was longer. Since the real process in the mass transfer zone is much more complicated than a set of arithmetical equations can describe, this correlation was derived for a rough estimation and to allow an understanding of the relation between breakthrough curve and operative conditions.

*CONCLUSIONS*

Adsorption behaviors of two phenolic compounds adsorbed onto two different types of adsorbent, XAD4 and F400, were investigated in this study. The solute partition coefficients ranged from 200 to 5,000. The Biot numbers varied from 2 to 30 indicating different experimental conditions. Concentrations and flow rates were in the range that is usually encountered in wastewater treatment. The effects of those operating parameters are summarized in Table 8.5. Agreement between model predictions and the experiments was good, suggesting the validity of the theory and the applicability of the combined diffusion

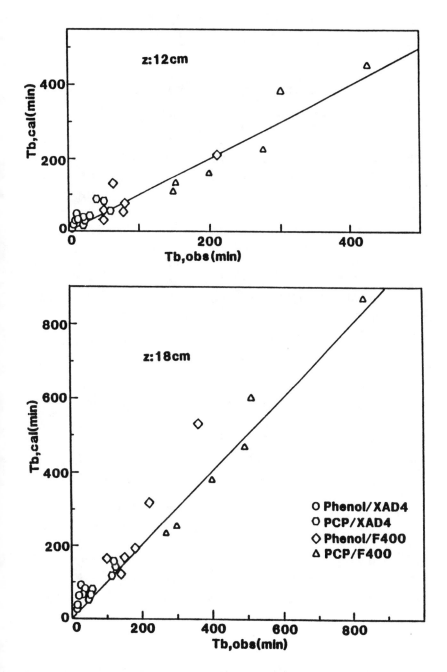

**Figure 8.13.**    Comparison between the Observed and Calculated
Breaktime

**Table 8.4.** Prediction of Breaktime in Single Component Systems

| Run | $(dX/dt)^{\dagger}$ @$X = 0.5$ | $T^*_{mtz}$ (min) | $L$ (cm) | $T_{b,cal}$ (min) | $T_{b,exp}$ (min) | $L$ (cm) | $T_{b,cal}$ (min) | $T_{b,exp}$ (min) |
|-----|------|------|------|------|------|------|------|------|
| A1 | 1.87 | 89 | 12.0 | 30 | 17 | 18.0 | 58 | 40 |
| A2 | 1.74 | 96 | 11.5 | 29 | 12 | 17.5 | 56 | 23 |
| A3 | 1.06 | 157 | 12.0 | 23 | 20 | 18.0 | 64 | 45 |
| A4 | 1.98 | 84 | 12.0 | 5 | 5 | 18.0 | 21 | 16 |
| A5 | 3.79 | 44 | 11.5 | 17 | 8 | 17.5 | 32 | 13 |
| A6 | 1.81 | 92 | 11.0 | 49 | 15 | 17.0 | 62 | 32 |
| B1 | 0.99 | 169 | 11.5 | 78 | 43 | 17.5 | 153 | 140 |
| B2 | 0.85 | 195 | 12.0 | 17 | 20 | 18.0 | 47 | 55 |
| B3 | 0.85 | 196 | 12.0 | 66 | 52 | 18.0 | 137 | 122 |
| B4 | 2.19 | 76 | 12.0 | 28 | 22 | 18.0 | 58 | 55 |
| B5 | 1.75 | 95 | 12.0 | 39 | 30 | 18.0 | 73 | 60 |
| B6 | 0.78 | 215 | 11.6 | 43 | 58 | 17.6 | 109 | 118 |
| C1 | 0.35 | 472 | 12.0 | 55 | 79 | 18.0 | 174 | 156 |
| C2 | 0.27 | 613 | 11.5 | 127 | 66 | 17.5 | 308 | 270 |
| C3 | 0.29 | 572 | 12.0 | 30 | 49 | 18.0 | 117 | 133 |
| C4 | 0.13 | 1269 | 12.0 | 180 | 205 | 18.0 | 494 | 372 |
| C5 | 0.43 | 388 | 12.0 | 61 | 53 | 18.0 | 162 | 106 |
| C6 | 0.36 | 462 | 12.0 | 72 | 80 | 18.0 | 194 | 185 |
| D1 | 0.39 | 430 | 12.0 | 115 | 146 | 18.0 | 261 | 310 |
| D2 | 0.21 | 805 | 12.0 | 146 | 192 | 18.0 | 366 | 390 |
| D3 | 0.13 | 1294 | 12.0 | 426 | 425 | 18.0 | 892 | 843 |
| D4 | 0.23 | 721 | 12.0 | 224 | 265 | 18.0 | 475 | 495 |
| D5 | 0.97 | 171 | 12.0 | 381 | 310 | 18.0 | 601 | 513 |
| D6 | 0.44 | 381 | 12.0 | 114 | 140 | 18.0 | 253 | 260 |

† Unit is $10^{-4}$ [sec$^{-1}$].

* The time for the mass transfer zone passing a fixed point in the bed is defined as:

$$T_{mtz} = \frac{1}{(dX/dt)_{X=0.5}}$$

model.

For microporous adsorbents with pore widths of the order of the size of the adsorbate molecules, surface diffusion was shown to be a possible mechanism of intraparticle diffusion because the pore space is not large enough for a liquid-filled phase. Therefore, the molecules contained in the pore are adsorbed molecules. However, the contribution of pore diffusion to the intraparticle diffusion in macroporous solids is appreciable when adsorption is weak and the pores contain a significant amount of adsorbate molecules in the liquid phase. Although the solid diffusion model could be applied to some adsorption systems as well, the combined diffusion model was more flexible in describing different adsorption behavior. Studies of multicomponent adsorption kinetics also showed that inclusion of a pore diffusion parameter could compensate for defects in the solid diffusion model. Therefore, the combined diffusion model will be developed further for multicomponent fixed-bed adsorption systems (Chapter X).

**Table 8.5.** Effects of Operative Conditions in Fixed Bed Adsorber – Single Component System

| Factor | Kinetic phenomenon | Reason |
|---|---|---|
| Column length $(L)$ | The curve is incomplete. The curve is steeper when $L$ is shorter. | $L < L_{mtz}$. The constant pattern is not formed yet. |
| Concentration $(C_0)$ | The curve appears earlier when $C_0$ is higher. | The solute distribution parameter is smaller, so that $V_{mtz}$ is faster. |
| | The curve is steeper when $C_0$ is higher. | The surface diffusivity is larger because $D_s = f(C_0)$, so that $L_{mtz}$ is shorter. |
| Interstitial velocity $(V)$ | The curve appears earlier when $V$ is higher. | The zone velocity is faster because $V_{mtz} = f(V)$. |
| | The curve is steeper when $V$ is higher. | The film mass transfer coefficient is larger because $K_f = f(V)$, so that $L_{mtz}$ is shorter. |
| Particle size $(R_p)$ | The curve is steeper when $R_p$ is smaller. | The cumulative rate increases when $R_p$ is smaller, so that $L_{mtz}$ is reduced. |
| Adsorption system (Sorbate-Sorbent) | The curve appears earlier when the system has a weaker adsorptive capability. | The zone velocity is faster due to a smaller solute distribution parameter. |
| | A steeper curve is generated from a weaker adsorption system. | Adsorption zone is shorter due to smaller solute distribution parameter and/or larger intraparticle diffusion coefficients. |

*REFERENCES*

1. Rosen, J.B. *Ind. Eng. Chem.*, **46**, 1590 (1954).
2. Stuart, F.X., and D.T. Camp, *AIChE. Symp.*, **69**, 33 (1973).
3. Crittenden, J.C., and W.J. Weber, Jr., *ASCE. EE.*, **104**, 185 and 433 (1978).
4. Cooper, R.S., and D.A. Liberman, *Ind. Eng. Chem. Fundam.*, **9**, 620 (1970).
5. Rasmuson, A. *AIChE. J.*, **27**, 1032 (1981).
6. McKay, G. *Chem. Eng. Res. Des.*, **62**, 235 (1984).
7. Neretnieks, I. *Chem. Eng. Sci.*, **31**, 465 and 1029 (1976).
8. Furusawa, T., and J.M. Smith, *Ind. Eng. Chem. Fundam.*, **12**, 197 (1973).
9. Fleck, R.D., Jr., D.J. Kirwan, and K.R. Hall, *Ind. Eng. Chem. Fundam.*, **12**, 95 (1973).
10. Weber, T.W., and R.K. Chakravorti, *AIChE. J.*, **20**, 228 (1974).
11. Merk, W., W. Fritz, and E.U. Schlünder, *Chem. Eng. Sci.*, **36**, 743 (1981).
12. Garg, D.R., and D.M. Ruthven, *Chem. Eng. Sci.*, **28**, 791 (1973).
13. Weber, W.J., Jr., and J.C. Crittenden, *J. WPCF.*, **47**, 924 (1975).
14. Hashimoto, K., K. Miura, and M. Tsukano, *J. Chem. Eng. Japan*, **10**, 27 (1977).
15. Villadsen, J.V., and W.E. Stewart, *Chem. Eng. Sci.*, **22**, 1483 (1967).
16. Finlayson, B.A. *Chem. Eng. Sci.*, **26**, 1081 (1971).
17. Liapis, A.I., and D.W.T. Rippin, *Chem. Eng. Sci.*, **33**, 593 (1978).
18. Crittenden, J.C., B.W.C. Wong, W.E. Thacker, V.L. Snoeyink, and R.L. Hinrichs, *J. WPCF.*, **52**, 2780 (1980).
19. Raghavan, N.S., and D.M. Ruthven, *AIChE. J.*, **29**, 922 (1983).
20. Liu, K.T., and W.J. Weber, Jr., *J. WPCF.*, **53**, 1541 (1981).
21. Rasmuson, A., and I. Neretnieks, *AIChE. J.*, **26**, 686 (1980).
22. Hindmarsh, A.C. *GEAR: Ordinary Differential Equation System Solver, UCID-30001 Rev. 3.* Lawrence Livermore Laboratory, P.O.Box 808, Livermore, CA (1974).
23. Hand, D.W., J.C. Crittenden, and W.E. Thacker, *ASCE. EE.*, **110**, 440 (1984).
24. Vortmeyer, D., and K. Michael, *Chem. Eng. Sci.*, **40**, 2135 (1985).
25. Cover, A.E., and C.D. Wood, *Appraisal of Granular Carbon Contacting: Phase III. Engineering Design and Cost Estimate of Granular Carbon Tertiary Waste Water Treatment Plant.* Taft Water Research Center Report No. TWRC-12, Cincinnati, OH (1969).
26. Fritz, W., W. Merk, and E.U. Schlünder, *Chem. Eng. Sci.*, **36**, 731

(1981).

27. Rasmuson, A. *Chem. Eng. Sci.*, **40**, 621 (1985).
28. Asai, S., Y. Konish, and H. Maeda, *Chem. Eng. Sci.*, **40**, 1573 (1985).
29. Moharir, A.S., D. Kunzru, and D.N. Saraf, *Chem. Eng. Sci.*, **35**, 1795 (1980).
30. Wakao, N., and T. Funazkri, *Chem. Eng. Sci.*, **33**, 1375 (1978).
31. Coulson, J.M., and J.F. Richardson, *Chemical Engineering.* 2nd ed., Pergamon Press, Oxford, New York (1979).
32. Colwell, C.J., and J.S. Dranoff, *Ind. Eng. Chem. Fundam.*, **10**, 65 (1971).

# CHAPTER IX

## FIXED BED MODELING AND DESIGN FOR GAS ADSORPTION SYSTEMS

Contributing Author: M.C. Yeh

## INTRODUCTION

Adsorption systems for solvent recovery or air purification generally employ a fixed bed adsorber. The design of an adsorber requires information on the dynamics of the system. The three main parameters involved in design are the adsorption isotherm, the external mass transfer coefficient and the intraparticle mass transfer coefficient.

To describe the adsorption process in the fixed bed, two theoretical models, the effective diffusion model [1-3] and the combined diffusion model [4-6], were developed and are compared in this chapter. Adsorption parameters for different adsorbate-adsorbent systems were determined in laboratory experiments and used to test the general applicability of the models.

## THEORY AND NUMERICAL ANALYSIS

The design models for fixed bed adsorbers require information on: external mass transfer resistance, intraparticle diffusion resistance and local equilibrium. The assumptions used in the models are: (1) negligible pressure drop, (2) constant hydraulic loading, (3) negligible axial dispersion, (4) isothermal conditions, and (5) constant surface diffusion coefficient.

### Combined Diffusion Model

If the porous adsorbent is considered to be a heterogeneous system, the internal diffusion can be expressed by the two possible simultaneous mechanisms: molecular Knudsen diffusion and surface diffusion. The material balance in a particle can be described by Equations (2.46)–(2.49). The dimensionless form of Equation (2.46) is shown as Equation (9.1).

$$\frac{\partial \overline{C}_p}{\partial \overline{T}} = \left(\frac{1}{n}\overline{C}_p^{1/n-1}\right)^{-1}\frac{D_g}{G}\left[\left(1+\beta\overline{C}_p^{1/n-1}\right)\frac{1}{\overline{r}^2}\frac{\partial}{\partial \overline{r}}\left(\overline{r}^2\frac{\partial \overline{C}_p}{\partial \overline{r}}\right)\right.$$
$$\left. + \left(\frac{1}{n}-1\right)\beta\left(\frac{\partial \overline{C}_p}{\partial \overline{r}}\right)^2\right] \tag{9.1}$$

The coupled initial and boundary conditions are:

$$\overline{C}_p = 0 \qquad \text{when } \overline{T} = 0 \tag{9.2}$$

$$\partial \overline{C}_p/\partial \overline{r} = 0 \qquad \text{at } \overline{r} = 0 \tag{9.3}$$

$$St(\overline{C} - \overline{C}_s) = \frac{\partial}{\partial \overline{T}}\int_0^1 \overline{q}\,\overline{r}^2\,d\overline{r} \tag{9.4}$$

The dimensionless form of the PDE for the fluid phase is:

$$\frac{\partial \overline{C}}{\partial \overline{T}} = -D_g\frac{\partial \overline{C}}{\partial \overline{z}} - D_g\alpha(\overline{C} - \overline{C}_s) \tag{9.5}$$

The boundary and initial conditions are:

$$\overline{C} = 0 \qquad \text{when } \overline{T} = 0 \tag{9.6}$$

$$\partial \overline{C}/\partial \overline{T} = 0 \quad \text{and} \quad \overline{C} = 1.0 \qquad \text{at } \overline{z} = 0 \text{ when } \overline{T} > 0 \tag{9.7}$$

Dimensionless parameters are defined as follows:

$$G = \frac{VKC_0^{1/n-1}R_p^2}{L\epsilon_p D_p}; \qquad \beta = \frac{q_e D_s}{C_0\epsilon_p D_p n}$$

$$\alpha = \frac{1-\epsilon_b}{\epsilon_b}\frac{3}{R_p}\frac{L}{V}k_f; \qquad D_g = \frac{q_e(1-\epsilon_b)}{C_0\epsilon_b}$$

$$St = \frac{k_f\tau(1-\epsilon_b)}{R_p\epsilon_b}; \qquad \overline{T} = \frac{t}{D_g\tau}$$

$$\tau = \frac{L}{V}; \qquad \overline{C} = \frac{C}{C_0}$$

$$\overline{z} = \frac{z}{L}; \qquad \overline{q} = \frac{q}{Q_e}$$

To apply the orthogonal collocation method [7], Equations (9.1) and (9.5) are transformed as follows:

Calculation in particles:

$$\frac{d\overline{C}_p}{d\overline{T}}\Big|_{\overline{r}_i,\overline{z}_i} = \left(\frac{1}{n}\overline{C}_{p(\overline{r}_i)}^{1/n-1}\right)\frac{D_g}{G}\left[\left(1 + \beta\overline{C}_{p(\overline{r}_i)}^{1/n-1}\right)\sum_{j=1}^{Nc} B_{ij}\overline{C}_p(\overline{r}_j)\right.$$
$$\left. + \left(\frac{1}{n} - 1\right)\beta\left(\sum_{j=1}^{Nc} A_{ij}\overline{C}_p(\overline{r}_j)\right)^2\right] \tag{9.8}$$

Calculation in fluid phase:

$$\frac{d\overline{C}}{d\overline{T}}\Big|_{\overline{z}_i} = -D_g\sum_{j=1}^{Mc} A_{ij}\overline{C}(\overline{z}_j) - D_g\alpha\left(\overline{C}(\overline{z}_i) - \overline{C}_s(\overline{z}_i)\right) \tag{9.9}$$

Calculation at interphase:

$$\frac{d\overline{q}}{d\overline{T}}\Big|_{\overline{r}_{Nc}} = \left[St\left(\overline{C}(\overline{z}_i) - \overline{C}_s(\overline{z}_i)\right) - \sum_{i=1}^{Nc-1} W_i\frac{d\overline{q}(\overline{r}_i)}{d\overline{T}}\right]/W_{Nc} \tag{9.10}$$

where $A_{ij}$, $B_{ij}$ = collocation coefficients are determined according to system and collocation number $Nc$, $W_i$ = weighting coefficients and $\overline{r}_i$ = interior collocation point.

*Effective Diffusion Model*

This model describes the intraparticle diffusion in terms of a single parameter called effective intraparticle diffusion coefficient, $D_e$. A detailed theoretical description of this kinetic parameter is given in Chapter IV. A material balance for a spherical particle results in the following dimensionless diffusion equation:

$$\frac{\partial\overline{q}}{\partial\overline{T}} = \frac{Ed}{\overline{r}^2}\frac{\partial}{\partial\overline{r}}\left(\overline{r}^2\frac{\partial\overline{q}}{\partial\overline{r}}\right) \tag{9.11}$$

where $Ed = D_eD_g\tau/R_p^2$.
Equivalent boundary and initial conditions are:

$$\overline{q} = 0 \qquad \text{when } \overline{T} = 0 \tag{9.12}$$

$$\partial\overline{q}/\partial\overline{r} = 0 \qquad \text{at } \overline{r} = 0 \tag{9.13}$$

$$St(\overline{C} - \overline{C}_s) = \frac{\partial}{\partial \overline{T}} \int_0^1 \overline{q}\,\overline{r}^2 d\overline{r} \qquad \text{at } \overline{r} = 1 \qquad (9.14)$$

The gas-phase material balance in a fixed bed has the following dimensionless form:

$$\frac{\partial \overline{C}}{\partial \overline{T}} = -D_g \frac{\partial \overline{C}}{\partial \overline{z}} - 3D_g St(\overline{C} - \overline{C}_s) \qquad (9.15)$$

The boundary and initial conditions are:

$$\overline{C} = 0 \qquad \text{when } \overline{T} = 0 \qquad (9.16)$$

$$\overline{C} = 1.0 \quad \text{and} \quad \partial \overline{C}/\partial \overline{T} = 0 \qquad \text{at } \overline{z} = 0 \qquad (9.17)$$

To be solved by orthogonal collocation method, the following equations are derived as:
Solid phase:

$$\frac{d\overline{q}}{d\overline{T}}\Big|_{\overline{r}_i, \overline{z}_i} = Ed \sum_{j=1}^{Nc} B_{ij} \overline{q}(\overline{r}_j) \qquad (9.18)$$

Fluid phase:

$$\frac{d\overline{C}}{d\overline{T}}\Big|_{\overline{z}_i} = -D_g \sum_{j=1}^{Mc} A_{ij} \overline{C}(\overline{z}_j) - 3D_g St(\overline{C}(\overline{z}_i) - \overline{C}_s(\overline{z}_i)) \qquad (9.19)$$

Particle surface:

$$\frac{d\overline{q}}{d\overline{T}} = \left[ St(\overline{C}(\overline{z}_i) - \overline{C}_s(\overline{z}_i)) - \sum_{i=1}^{Nc-1} W_i \frac{d\overline{q}(\overline{r}_i)}{d\overline{T}} \right] / W_{Nc} \qquad (9.20)$$

## EXPERIMENTAL RESULTS AND DISCUSSION

### Experimental Procedure

The two adsorbents used in this study are beaded activated carbon BAC and resin XAD-4. The six adsorbates: toluene, xylene, benzene, carbon tetrachloride, tetrachloroethylene, and trichloroethylene, were

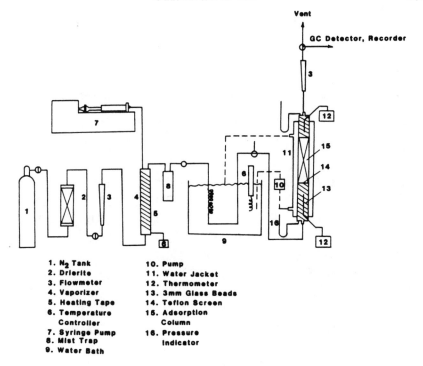

**Figure 9.1.**    Schematic Diagram of Experimental Setup

carried by nitrogen in the fixed-bed experiments. The properties of adsorbents and adsorbates were described in Chapter IV.

Adsorption breakthrough curves were measured with the apparatus shown in Figure 9.1. The flow of nitrogen-carried gas, drawn from a high-pressure gas cylinder, was controlled with a regulator and needle valves and measured with a calibrated flowmeter. To incorporate the required concentration of adsorbate vapor, the model organic compound in a gas tight syringe was introduced into the vaporizer using a Sage Instrument syringe pump. Prior to entering the vaporizer, the nitrogen was passed through a column of drierite for humidity removal and was monitored by means of thermocouples (Cole Parmer 700 series). A U tube-type pressure gauge was used to measure the pressure drop through the bed.

The experimental data were obtained using a chromatography (GOW-MAC 750), detecting the organic vapor by a flame ionization detector. The range of operating conditions in this study were: concentration $C_0$ from 400 ppm to 4000 ppm, packing height from 5 cm to 35 cm, superficial velocity from 10 cm/sec to 50 cm/sec. Since the ratio of the column diameter to the particle diameter, $D_{col}/d_p$, was larger than 35, and the ratio of the column length to the particle

**Table 9.1.**    Physical Parameters of Fixed Bed Adsorbers

| Run No. | System | $C_0$ (ppm) | $\epsilon_b$ (-) | $V$ (cm/sec) | $L$ (cm) |
|---------|--------|-------------|------------------|--------------|----------|
| R1  | $C_7H_8$/XAD-4        | 480  | 0.46 | 15.5 | 15.3 |
| R2  | $C_7H_8$/XAD-4        | 1200 | 0.45 | 15.7 | 25.0 |
| R3  | $C_7H_8$/XAD-4        | 1165 | 0.45 | 15.7 | 7.5  |
| R4  | $C_7H_8$/XAD-4        | 4870 | 0.46 | 15.7 | 19.7 |
| R5  | $C_7H_8$/XAD-4        | 4870 | 0.47 | 40.0 | 25.3 |
| R6  | $C_6H_6$/XAD-4        | 1450 | 0.45 | 15.7 | 18.8 |
| R7  | $C_6H_6$/XAD-4        | 1400 | 0.45 | 32.7 | 19.1 |
| R8  | $C_6H_6$/XAD-4        | 6160 | 0.47 | 15.3 | 17.1 |
| R9  | $p$-$C_8H_{10}$/XAD-4 | 1990 | 0.46 | 16.0 | 7.5  |
| R10 | $p$-$C_8H_{10}$/XAD-4 | 4320 | 0.45 | 15.7 | 12.3 |
| R11 | $CCl_4$/XAD-4         | 1270 | 0.45 | 31.7 | 17.4 |
| R12 | $C_2HCl_3$/XAD-4      | 1525 | 0.46 | 15.9 | 14.0 |
| R13 | $C_2HCl_3$/XAD-4      | 1500 | 0.45 | 30.8 | 19.7 |
| R14 | $C_2HCl_3$/XAD-4      | 5840 | 0.46 | 15.9 | 18.2 |
| R15 | $C_2Cl_4$/XAD-4       | 1340 | 0.47 | 15.9 | 14.5 |
| R16 | $C_2Cl_4$/XAD-4       | 1300 | 0.46 | 32.1 | 13.0 |
| R17 | $C_2Cl_4$/XAD-4       | 5000 | 0.46 | 16.4 | 17.3 |
| C1  | $C_6H_6$/BAC          | 1500 | 0.63 | 15.4 | 7.2  |
| C2  | $C_7H_8$/BAC          | 1200 | 0.63 | 15.7 | 10.0 |
| C3  | $C_7H_8$/BAC          | 1236 | 0.62 | 15.6 | 7.3  |
| C4  | $p$-$C_8H_{10}$/BAC   | 1075 | 0.63 | 15.4 | 4.3  |
| C5  | $CCl_4$/BAC           | 1428 | 0.63 | 14.8 | 5.3  |
| C6  | $C_2HCl_3$/BAC        | 1525 | 0.64 | 15.0 | 6.0  |

diameter, $L/d_p$, was larger than 75, the axial dispersion effect was assumed negligible [8,9]. All experiments were run at a temperature of $25\pm0.5°C$. Table 9.1 lists the operating conditions of some typical runs.

Due to the adsorption heat for the gas-solid systems, the isothermal condition is not attainable as easily as in the liquid-solid systems [10]. Figure 9.2 shows the breakthrough curve and temperature history at

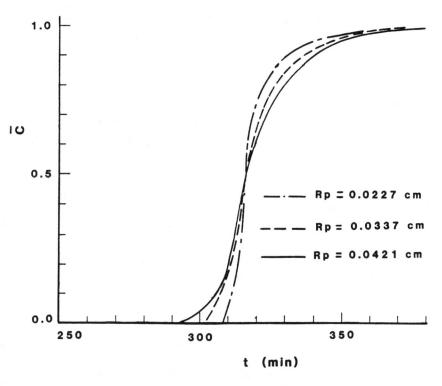

**Figure 9.2.**    Effluent Concentration Profile of $C_7H_8$ from XAD-4 Bed

the end of the bed. The thermal wave leads the breakthrough curve, with the maximum temperature rise occuring at about $C/C_0 = 0.5$. The maximum temperature rise was about 0.5°C, which was assumed to be negligible because this variance was relatively small when the operating temperature was 25°C.

*Sensitivity Analysis*

To test the effect of a specific parameter on the model results, the value of that parameter was held constant at its normal average value while other parameters were varied over their normal operating ranges. An example of this type of analysis will be shown for a toluene/BAC adsorption system, isotherm parameters are provided in Table 4.2. The values for the fixed-bed parameters, particle size $R_p$, superficial velocity $V$, influent vapor concentration $C_0$, and bed length $L$, were as follows: $R_p = 0.0337$ cm, $V = 15.7$ cm/sec, $C_0 = 1200$ ppm and $L =$

10 cm. The parameters varied were the surface diffusion coefficient $D_e$, the particle size $R_p$ and the external mass transfer coefficient $k_f$. The results of the sensitivity analysis are as follows:

1. Collocation numbers: It was observed that if the collocation points for the external fluid phase $(M_c)$ is greater than eight and the number of interior collocation points $(N_c)$ is greater than seven, the effect of the collocation numbers is insignificant. In this study, therefore, eight collocation points were used for the external field and seven points for the concentration profile through an individual particle.

2. Particle size: The particle size and shape affects the pressure drop through the adsorbent bed and the diffusion rate into the particle. Figure 9.3 shows the evidence that the breakthrough time increases as the particle size decreases. This phenomenon involves two parameters: the mass transfer coefficient and the outside surface area per unit particle volume. As the particle size increases, the gas film resistance and the interior diffusion path increase.

3. Effective diffusion coefficient: If a larger $D_e$ is used, a sharper breakthrough curve can be observed, as shown in Figure 9.4. Larger $D_e$ values indicate lower interior diffusion resistance; therefore, the length of mass transfer zone is reduced.

4. External mass transfer coefficient: The $k_f$ values for gas-solid systems were estimated by using Equation (2.26). As shown in Figure 9.5, the breakthrough point is delayed and the gradient of the curve becomes steeper, when a larger $k_f$ value is used. However, the effect of this parameter on the model is moderate.

*Modeling Comparison*

Table 9.2 contains the effective diffusion coefficient, surface diffusion coefficient, pore diffusion coefficient and the external mass transfer coefficient for each experimental adsorption system. The effective diffusion coefficient, as mentioned in Chapters IV and V, is a global effective value obtained by using a differential reactor technique. Although results showed that the $D_e$ values are concentration dependent, the diffusivities are the average values defined by the evaluation procedure. This means that the measured diffusivity is an averaged value for the adsorption process.

The surface diffusion coefficient in the combined diffusion model is an effective value for a specified external concentration, while the pore diffusion coefficient is a constant (Chapter VI). Since the effective diffusion coefficients were estimated over a finite concentration change,

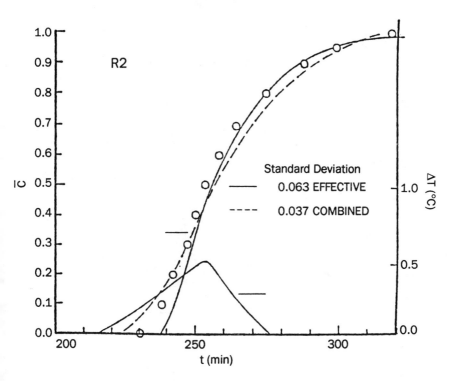

**Figure 9.3.** Effect of Particle Size on Predicted Breakthrough Curve

it is not proper to correlate the diffusivities as:

$$D_e = f(C) \quad \text{\tiny Sthe} \tag{9.21}$$

Figure 9.6 shows that the model using correlation Equation (9.21) does not fit the fixed-bed rate data. The shape of the predicted breakthrough curve is not as steep as the experimental curve. This implies that the effective diffusivity related to each concentration may be underestimated, as described in Chapter III.

The shape of the breakthrough curve and the breakthrough time may also be affected by different column lengths, flow rate, and input concentrations. Figure 9.7 shows a set of curves generated for different bed depths. The breakthrough time (taken as the first observed toluene effluent) was found to be linear with bed length, as shown in Figure 9.8.

Figures 9.9 and 9.10 show the small observed effect of gas velocity on the shape of the breakthrough curve. By increasing the gas

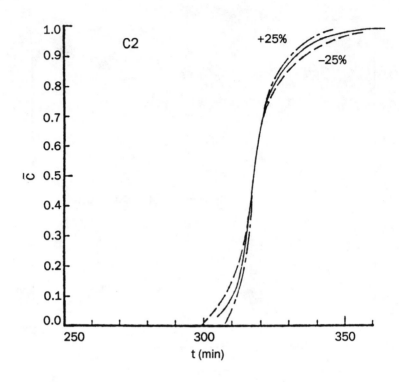

**Figure 9.4.**    Effect of Effective Diffusivity on Predicted Breakthrough Curve

velocity by a factor of 2.5, the shape of the breakthrough curve was unchanged, but the breakpoint appeared much earlier. The effect on the temperature was found to be small with the maximum temperature rise slightly higher for the higher velocity run.

Figures 9.2 and 9.9 also allow a comparison of experimental data to model predictions. Agreement between the theoretical and experimental curve is fairly good, particularly in view of the fact that all the parameters were estimated from the differential reactor adsorption studies. The small discrepancies between theory and experiments are well within the margin of error, arising from uncertainty in the parameters. In particular, the values of the equilibrium constants $K$ and $n$ are subject to an estimated uncertainty $\pm 10\%$.

When comparing predicted breakthrough curves with experimentally measured breakthrough curves, the standard deviation of the combined diffusion model and the effective model are similar. The residuals between the two model predictions and the experimental results versus experimental time are shown in Figures 9.11 and 9.12.

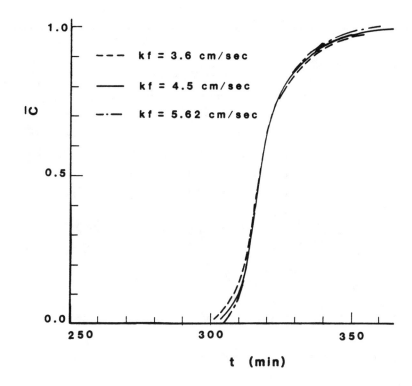

**Figure 9.5.** Effect of Gas Film Resistance on Predicted Breakthrough Curve

It is observed that at breakthrough time, the residuals of the combined diffusion model are always smaller than those of the effective diffusion model. Figure 9.13 also shows the theoretically predicted times versus the observed times for $C/C_0 = 0.05$. The predicted times for the combined diffusion model are always slightly smaller than those for the effective diffusion model.

A further series of experiments was performed to test the applicabilities of the models to predict the breakthrough curves for more favorable adsorption systems, i.e., the adsorption of organic vapor on activated carbon. The experimental and theoretical breakthrough curves are compared in Figures 9.14-9.18. Since the contribution of the pore diffusion to the total mass transfer is less than 10% in carbon adsorption, the effective diffusion coefficients used for the effective diffusion model are very close to the surface diffusion coefficients used for the combined diffusion model, as shown in Table 9.2. Figures 9.14-9.18 also show the standard deviations between the experimental

**Table 9.2.** Mass Transfer Parameters for Fixed Bed Adsorbers

| Run No. | System | $D_e$ $\times 10^8$ | $D_s$ $\times 10^8$ | $D_p$ $\times 10^4$ | $k_f$ |
|---------|--------|------|------|------|------|
| R1  | $C_7H_8$/XAD-4      | 9.5  | 7.7  | 11.1 | 4.5 |
| R2  | $C_7H_8$/XAD-4      | 15.5 | 12.5 | 11.1 | 4.5 |
| R3  | $C_7H_8$/XAD-4      | 14.7 |      |      | 4.5 |
| R4  | $C_7H_8$/XAD-4      | 38.0 | 29.0 | 11.1 | 4.5 |
| R5  | $C_7H_8$/XAD-4      | 38.0 | 29.0 | 11.1 | 7.2 |
| R6  | $C_6H_6$/XAD-4      | 22.5 | 14.6 | 10.7 | 4.7 |
| R7  | $C_6H_6$/XAD-4      | 20.8 | 12.4 | 10.7 | 6.6 |
| R8  | $C_6H_8$/XAD-4      | 56.3 | 36.6 | 10.7 | 4.7 |
| R9  | $p$-$C_8H_{10}$/XAD-4 | 13.5 | 10.4 | 10.1 | 4.7 |
| R10 | $p$-$C_8H_{10}$/XAD-4 | 15.3 | 11.8 | 10.1 | 4.4 |
| R11 | $CCl_4$/XAD-4       | 26.4 | 17.4 | 9.0  | 6.0 |
| R12 | $C_2HCl_3$/XAD-4    | 27.0 | 21.0 | 9.7  | 4.4 |
| R13 | $C_2HCl_3$/XAD-4    | 25.2 | 18.9 | 9.7  | 6.2 |
| R14 | $C_2HCl_3$/XAD-4    | 61.1 | 45.1 | 9.7  | 4.4 |
| R15 | $C_2Cl_4$/XAD-4     | 10.1 | 7.8  | 8.1  | 4.3 |
| R16 | $C_2Cl_4$/XAD-4     | 9.5  | 7.4  | 8.1  | 6.2 |
| R17 | $C_2Cl_4$/XAD-4     | 19.8 | 15.0 | 8.1  | 4.5 |
| C1  | $C_6H_6$/BAC        | 17.7 | 16.5 | 14.4 | 4.7 |
| C2  | $C_7H_8$/BAC        | 17.7 |      |      | 4.5 |
| C3  | $C_7H_8$/BAC        | 21.8 | 20.9 | 12.5 | 4.5 |
| C4  | $p$-$C_8H_{10}$/BAC | 18.4 | 14.7 | 12.9 | 4.4 |
| C5  | $CCl_4$/BAC         | 28.8 | 27.4 | 9.9  | 4.6 |
| C6  | $C_2HCl_3$/BAC      | 17.8 | 16.4 | 10.0 | 4.2 |

Units in this table are $D_e$: [cm$^2$/sec], $D_s$: [cm$^2$/sec], $D_p$: [cm$^2$/sec], $k_f$: [cm/sec].

data and the simulated curves using these two models. It is evident that there is a fairly good agreement between the experimental and theoretical curves.

*CONCLUSIONS*

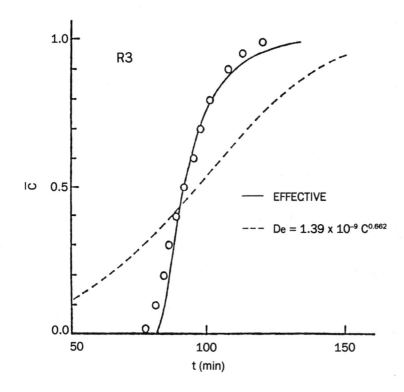

**Figure 9.6.**    Effect of the Concentration Dependent Diffusion Coefficient on the Breakthrough Curve

The fixed bed studies covered the expected values encountered in the adsorber column operation. The favorable sorption isotherms produced sharp breakthrough curves, which are relatively unaffected by moderate fluctuations in the input variables. The breakthrough time is shortened and the breakthrough curve is sharpened when the influent concentration is larger. The effect of gas velocity on the shape of the breakthrough curve is negligible. This phenomenon implies that the intraparticle mass transfer resistance is more important than that of the external film resistance.

When comparing the two models for a fixed bed adsorber, the breakthrough time is delayed for the effective diffusion model. The standard deviation between the experimental data and the predicted values of the effective diffusion model, is a little higher than that of the combined diffusion model for the adsorption of organic vapors on XAD-4. In the activated carbon adsorption system, the surface diffusion

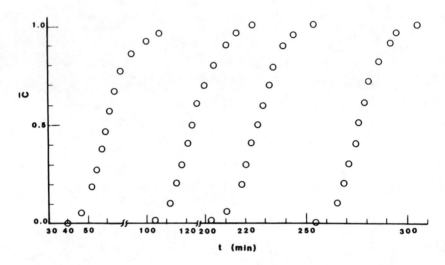

**Figure 9.7.**    Breakthrough Curves for Different Bed Depths

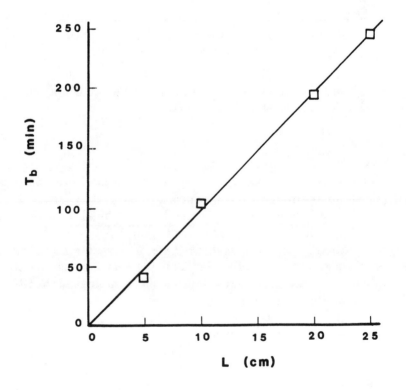

**Figure 9.8.**    Effect of Bed Length on the Breakthrough Time

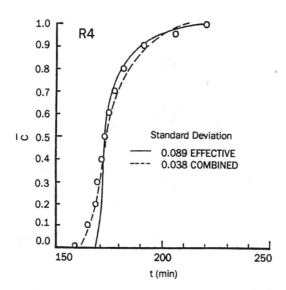

**Figure 9.9.** Effluent Concentration Profile of $C_7H_8$/XAD-4

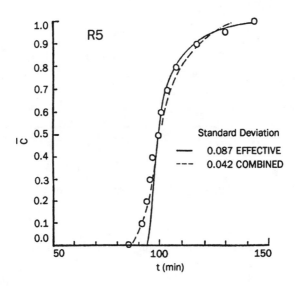

**Figure 9.10.** Effluent Concentration Profile of $C_7H_8$/XAD-4

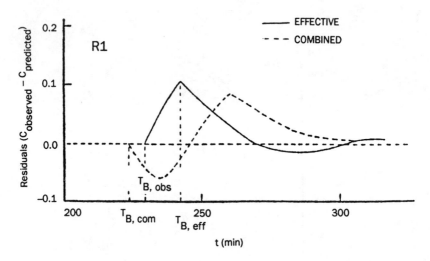

**Figure 9.11.**   Residuals versus Time for Run R1

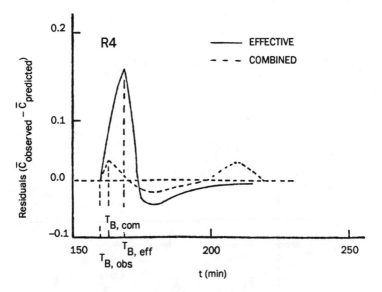

**Figure 9.12.**   Residuals versus Time for Run R4

coefficient for the combined diffusion model is close to the effective diffusivity for the effective diffusion model. Therefore, both models fit the experimental data well.

The equilibrium and kinetic parameters used in the fixed bed model predictions were all determined from a differential bed reactor. This implies that the kinetic behavior in a differential reactor is very similar

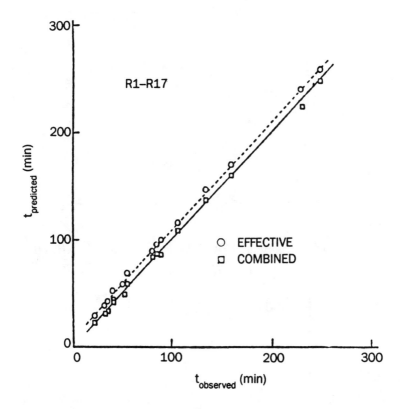

**Figure 9.13.** Theoretically Predicted Time versus Experimentally Observed Time at $C/C_0 = 0.05$

to that in the fixed bed column, and the differential bed reactor is a convenient method to development of adsorption parameters needed for long column design.

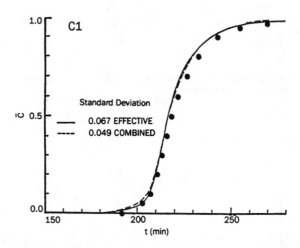

**Figure 9.14.** Effluent Concentration Profile of $C_6H_6$/BAC

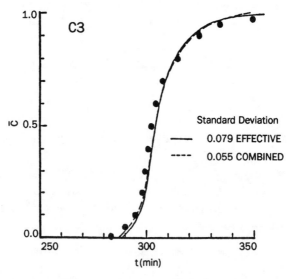

**Figure 9.15.** Effluent Concentration Profile of $C_7H_8$/BAC

*REFERENCES*

1. Aguwa, A.A. *Evaluation of Sorption Behavior of Single and Multicomponent Organics with Synthetic Resin.* Ph.D. Thesis, Illinois Institute of Technology, Chicago, IL (1984).
2. Sarlis, J.N. *Sorptive Behavior of Hazardous Organic Solutions by Direct Differential Reactor Measurements.* M.S. Thesis, Illinois In-

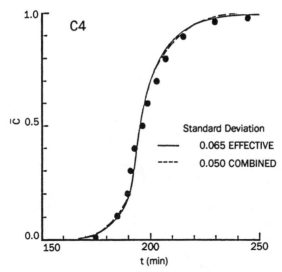

**Figure 9.16.**   Effluent Concentration Profile of $p$-$C_8H_{10}$/BAC

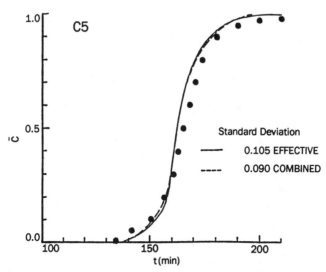

**Figure 9.17.**   Effluent Concentration Profile of $CCl_4$/BAC

stitute of Technology, Chicago, IL (1985).

3. Noll, K.E., and J.N. Sarlis, *J. APCA.*, **38**, 1512 (1988).
4. Costa, E., G. Calleja, and F. Domingo, *AIChE. J.*, **31**, 982 (1985).
5. Costa, C., and A. Rodrigues, *Chem. Eng. Sci.*, **40**, 983 (1985).
6. Leyva-Ramos, R., and C.J. Geankoplis, *Chem. Eng. Sci.*, **40**, 799 (1985).
7. Finlayson, B.A. *The Method of Weighted Residuals and Variational*

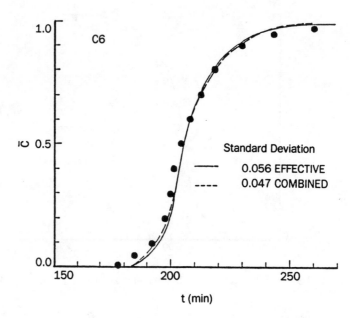

**Figure 9.18.**     Effluent Concentration Profile of $C_2HCl_3/BAC$

*Principles.* Academic Press, New York, NY (1972).
8. Thomas, W.J., and A.R. Qureshi, *Trans. Inst. Chem. Eng.*, **49**, 60 (1971).
9. Raghavan, N.S., and D.M. Ruthven, *AIChE. J.*, **29**, 922 (1983).
10. Sircar, S., R. Kumar, and K.J. Anselmo, *Ind. Eng. Proc. Des. Dev.*, **22**, 10 (1983).

# CHAPTER X

---

# MODELING OF MULTICOMPONENT
# WATER ADSORPTION SYSTEMS

---

## INTRODUCTION

The adsorption process as applied in practical water treatment situations usually involves multi-solute mixtures. Because of the competitive phenomenon during multi-solute adsorption, two factors have to be considered in the operation. First, there is the possibility that the most weakly adsorbed solute (therefore, the first to come off the column) will be toxic and will threaten the water users or receiving stream. Second, it may be necessary to remove the treatment system from service for regeneration prematurely, before the full service capacity of all the adsorbent is attained. This adds cost to an already expensive treatment process. A proper mathematical model enables consideration of this competitive effect in the design and monitoring of fixed bed adsorbers. Therefore, a modeling study of multicomponent fixed bed adsorption was conducted.

Several models presented in previous studies include the solid diffusion model [1-3], the pore diffusion model [4] and the combined diffusion model [3,5]. Because of the complexity of the mathematics, a linear driving force approximation was sometimes applied to solve specific problems [6-8]. Nevertheless, the models are usually extended from those applied to single component systems, and the competitive effect is accounted for through the multicomponent adsorption isotherm. Therefore, the information derived from single component systems can be used directly in the simulation of multicomponent adsorption. In some cases, however, diffusional interference is introduced into the diffusion model to interpret the phenomenon of interaction in the particle [3,8,9]. Since this interactive force is assumed in each pair of sorbates, the values of cross diffusion coefficients have to be evaluated from binary adsorption kinetics.

The model developed in this study is the combined solid and pore diffusion model. For comparison, the solid diffusion model was also applied. Both models are extended from the work in single component systems (Chapter VIII).

## EXPERIMENTAL RESULTS AND DISCUSSION

### Mathematical Model

The assumptions for the single component conditions are valid for each species that exists in a mixture. In the solution of the model it is assumed that $D_{mij} = 0$ and $D_{sij} = 0$ for $i \neq j$ implying that each species diffuses independently of the others. Nevertheless, the interactive and competitive effects during the adsorption process are accounted for through the equilibrium isotherm, because the assumption of local equilibrium holds at the interphase. Therefore, the model can be derived as follows.

Particle:

$$\frac{\partial Y_i}{\partial T} = \frac{N_{si}}{\bar{r}^2} \frac{\partial}{\partial \bar{r}} \left( \bar{r}^2 \frac{\partial Y_i}{\partial \bar{r}} \right) + \alpha_{pT} \frac{N_{pi}}{\bar{r}^2} \frac{\partial}{\partial \bar{r}} \left( \bar{r}^2 \frac{\partial X_{pi}}{\partial \bar{r}} \right) \tag{10.1}$$

Fixed bed:

$$\frac{\partial X_i}{\partial T} = -\frac{\partial X_i}{\partial \bar{z}} - \frac{1}{\alpha_{bT}} \frac{\partial Y_{av,i}}{\partial T} \tag{10.2}$$

where

$$Y_{av,i} = 3 \int_0^1 Y_i \bar{r}^2 \, d\bar{r} \tag{10.3}$$

Initial and boundary conditions:

$$\alpha_{bT} St_i (X_i - X_{si}) = N_{si} \frac{\partial Y_i}{\partial \bar{r}} + \alpha_{pT} N_{pi} \frac{\partial X_{pi}}{\partial \bar{r}} \qquad \text{at } \bar{r} = 1 \tag{10.4}$$

$$\alpha_{bT} St_i (X_i - X_{si}) = \frac{\partial}{\partial T} \int_0^1 Y_i \bar{r}^2 \, d\bar{r} \tag{10.5}$$

$$\partial Y_i / \partial \bar{r} = 0, \quad \partial X_{pi} / \partial \bar{r} = 0 \qquad \text{at } \bar{r} = 0 \tag{10.6}$$

$$Y_i = 0, \quad X_{pi} = 0, \quad X_i = 0 \qquad \text{when } T = 0 \tag{10.7}$$

$$X_i = X_{0i}, \quad \partial X_i / \partial T = 0 \qquad \text{at } \bar{z} = 0 \text{ when } T > 0 \tag{10.8}$$

Dimensionless groups and variables used in substitution are defined as follows.

$$Y_i = q_i / Q_{eT}; \quad Y_{av,i} = q_{av,i} / Q_{eT}$$

$$T = tV/L; \quad \bar{z} = z/L; \quad \bar{r} = r/R_p$$

$$X_i = C_i / C_{0T}; \quad X_{0i} = C_{0i} / C_{0T}$$

$$X_{p_i} = C_{p_i}/C_{0T}; \quad X_{si} = C_{si}/C_{0T}$$

$$\alpha_{pT} = C_{0T}\epsilon_p/Q_{eT}\rho_p; \quad \alpha_{bT} = C_{0T}\epsilon_b/Q_{eT}\rho_p(1-\epsilon_b)$$

$$N_{si} = D_{si}L/R_p^2 V; \quad N_{pi} = D_{p_i}L/R_p^2 V$$

$$St_i = k_{f_i}L(1-\epsilon_b)/R_p\epsilon_b V$$

where

$$C_{0T} = \sum_i^n C_{0i}; \quad Q_{eT} = \sum_i^n Q_{ei}$$

Since local equilibrium adjacent to the exterior surface of the particle and the interior pore walls is assumed, the relation of equilibrium at interphase is described by the IAS theory [10] in this study.

Normalized partial differential equations (PDE's) were solved by using orthogonal collocation method. By the principle of this numerical method, the PDE's for each species $i$ have to be transferred to be a set of first-order ordinary differential equations (ODE's), given as:

Intraparticle:

$$\left(\frac{dY_i}{dT}\right)_{jk} = N_{si}\left(\sum_{n=1}^{N+1} B_{jn}Y_{ink}\right) + \alpha_{pT}N_{pi}\left(\sum_{n=1}^{N+1} B_{jn}X_{p_{ink}}\right) \quad (10.9)$$

Fixed bed:

$$\left(\frac{dX_i}{dT}\right)_k = -\left(\sum_{m=1}^{M+2} A_{km}X_{im}\right) - 3St_i(X_{ik} - X_{sik}) \quad (10.10)$$

Particle surface:

$$\alpha_{bT}St_i(X_{ik} - X_{sik}) = \sum_{n=1}^{N+1} W_n\left(\frac{dY_i}{dT}\right)_{nk} \quad (10.11)$$

Figure 10.1 shows numerical analogue locations for the approximation of PDE's for the species $i$ in the fixed bed system. The program for this numerical method includes four major parts: (1) main program for the input, output and the definition of the normalized terms; (2) subroutine for the orthogonal collocation procedure; (3) subroutine for the calculation of equilibrium; and (4) GEAR subroutine [11] for solving ODEs. This program was written in Fortran IV language and runs on the Prime 500 computer system (Appendix).

*Experimental Procedure*

Two adsorbates used in this study were phenol and *p*-chlorophenol (reagent grade, Aldrich Chemical Company). The adsorbents were used Amberlite resin XAD-4 (Rhom and Haas Company) and Calgon's activated carbon F400 (Calgon Corporation). The properties and preparation of stock solutions and adsorbents, the structure of experimental column adsorption system, and the procedure of experiment are as described in Chapter III.

The UV-Spectrophotometer was used to measure liquid concentrations. In the range of dilute solution, the relation between absorbance and concentration follows Beer's law:

$$A_{uv} = K_\lambda C \qquad (10.12)$$

where $A_{uv}$ is the absorbance of ultraviolet light and $K_\lambda$ the extinction coefficient for a specified wavelength $\lambda$. The optimum wavelength is 269.5 nm for phenol and 279.7 nm for *p*-chlorophenol aqueous solutions. Concentrations in the mixtures were obtained according to the method introduced by Crittenden [12]:

$$A_{uv1} = K_{\lambda_{11}}C_1 + K_{\lambda_{12}}C_2 \qquad (10.13a)$$

$$A_{uv2} = K_{\lambda_{21}}C_1 + K_{\lambda_{22}}C_2 \qquad (10.13b)$$

where $K_{\lambda_{ij}}$ is the extinction coefficient for species $j$ at wavelength $\lambda_i$. The $K_{\lambda_{ij}}$'s of phenol are $1.41 \times 10^3$ abs-l/mol-cm (269.5 nm) and $6.16 \times 10^2$ abs-l/mol-cm (279.7 nm), and those of *p*-chlorophenol are $9.49 \times 10^2$ abs-l/mol-cm (269.5 nm) and $1.47 \times 10^3$ abs-l/mol-cm (279.7 nm). The standard deviation using Crittenden's method was about 5.8% using 54 prepared mixtures.

*Adsorption Equilibrium*

Competitive behavior is a characteristic phenomenon in multicomponent adsorption systems. Because the adsorption sites on the surface of the sorbent are limited, the adsorption of any adsorbate in a mixture is restricted by the coexistent species. A sorbate having strong affinity with the sorbent surface and weak affinity with the solvent usually represents a strongly adsorbed species; therefore, the

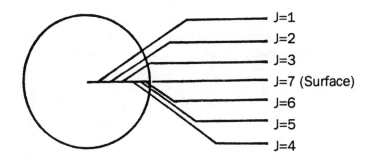

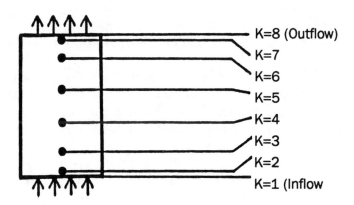

**Figure 10.1.**    Numerical Analogue Locations for the Approximation of PDE's in the Fixed-Bed System

sorbate has a high adsorption capacity. In a mixture, the strongly adsorbed species always has a greater tendency to occupy the adsorption sites than the weakly adsorbed species does. Therefore, the adsorption capacity of the weakly adsorbed species decreases more than the strongly adsorbed species does.

Experimental equilibrium data for the two binary adsorption systems, phenol+$p$-chlorophenol/XAD-4 and phenol+$p$-chlorophenol/F400, show that the sorbate, phenol, is a weakly adsorbed species while $p$-chlorophenol is a strongly adsorbed species. Therefore, phenol sorption is reduced more than $p$-chlorophenol at the same molar ratio. Although the adsorption of phenol and $p$-chlorophenol on F400 are much higher than on XAD-4, the reduced percentage of phenol adsorption in the mixture of phenol+$p$-chlorophenol/F400 is larger than that in the system of phenol+ $p$-chlorophenol/XAD-4. This is due to the higher selectivity of the adsorbent F400. This selectivity has a significant effect in the multicomponent fixed-bed adsorption studies.

The isotherm model applied to represent the binary adsorption equilibria is based on the IAS theory. It can be applied by using single component equilibrium isotherms and has been widely studied for different mixtures. The statistical results of applying the IAS model are listed in Table 10.1 and the deviations between experimental and predicted data are shown in Figures 10.2 and 10.3.

*Sensitivity Analysis*

Sensitivity analyses involving the effects of equilibrium parameters, rate parameters, and particle size were also carried out for a binary adsorption system. Some conclusions can be drawn from those results as follows:

1. The errors in both $K$- and $n$-values will affect the performance of the strongly adsorbed species more significantly than that of the weakly adsorbed species. This is because the weakly adsorbed species usually has a much lower capacity in the binary system, and the effect of isotherm parameters might be spread over two active zones. Furthermore, if a positive deviation in $K_1$- or $n_2$-value is employed in simulation, the weakly adsorbed species generates a delayed breakthrough curve and a smaller excess percentage, while the strongly adsorbed species gives an early appearance of breakthrough curve. This variation results from the relative capacities of sorbates in a mixture. Therefore, an increasing capacity of the weakly adsorbed species or decreasing amount of the strongly adsorbed species will produce the phenomena just mentioned.

**Table 10.1.**  Statistical Results of Equilibrium Studies

| Sorbate | Sorbent | $K_i$ | $n_i$ | $S_i$ | Data No. |
|---------|---------|-------|-------|-------|----------|
| Phenol(1) | XAD4 | 2.31 | 1.72 | 12.5 | 21 |
| $p$-Chlorophenol(2) | | 15.2 | 2.78 | 11.9 | |
| Phenol(1) | F400 | 86.7 | 5.46 | 18.5 | 21 |
| $p$-Chlorophenol(2) | | 138.0 | 5.96 | 9.3 | |

∗: Individual deviation is illustrated in Figures 10.2 and 10.3. Average deviation, $S_i$, for the solute $i$ in a binary system is defined as:

$$S_i = \sqrt{\frac{\sum_{j=1}^{m}\left(1 - \left(\frac{q_{i,cal}}{q_{i,exp}}\right)_j\right)^2}{m}} \times 100\%$$

2. The errors in both surface diffusion and pore diffusion coefficients have minor effects on the simulation of breakthrough curves. Results show that a positive deviation in $D_s$- or $D_p$-values of both sorbates will generate steeper breakthrough curves, because a larger diffusion coefficient means smaller diffusion resistance so that equilibrium state can be achieved more easily. If a positive variation of $D_s$- or $D_p$-value is tested for only one species, then this sorbate generates a steeper breakthrough curve and the other has a flatter one. This interactive effect occurs because one sorbate has a larger intraparticle diffusion flux and will resist the diffusion from the other, so that the mass transfer zone of the other sorbate will be stretched.

3. The effect of external mass transfer coefficient is usually minute, especially for the range of error near 20%. If this effect is considered individually, then the simulated breakthrough curves of both sorbates will rise when larger $k_f$-values are used. However, the effect is more sensitive to variations in $k_f$-values of the strongly adsorbed species, because its Biot number is much smaller than that of the weakly adsorbed species.

A modeling sensitivity analysis can provide some useful information

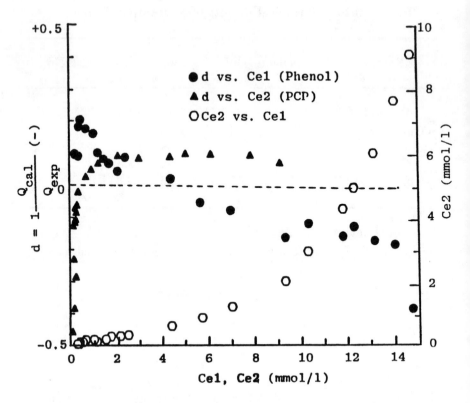

**Figure 10.2.** Comparison of IAS Model for Phenol(1)+
p-Chlorophenol(2)/XAD-4

to judge which parameters may cause serious deviation in the modeling simulation. In the case of multicomponent adsorption, however, this analysis is complicated, because a variance in the parameter of any sorbate will affect all species in the system. Analytic results based on an example of phenol(1)+p-chlorophenol(2)/XAD4 show that if an error of 10% is assumed on any parameter, then an average deviation from the predicted breakthrough curves was found to be 10% for equilibrium parameters, 7% for particle size, 4% for surface diffusion coefficients and insignificant for pore diffusion and film mass transfer coefficients. This suggests that the reliability of the isotherm model and the particle size estimation are important in the modeling simulation.

*Modeling Comparison*

The purpose of developing a model is to simulate the breakthrough

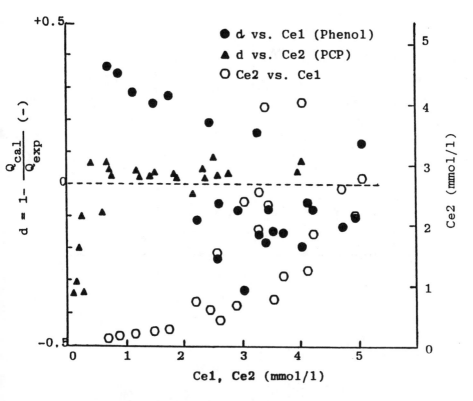

**Figure 10.3.** Comparison of IAS Model for Phenol(1)+
p-Chlorophenol(2)/F400

curves and characterize the excess amount of the weakly adsorbed
species. Consider a case of binary adsorption in a fixed bed, in which
the breakthrough curves of both sorbates at different column lengths
have the pattern as shown in Figure 10.4.

Usually, the weakly adsorbed species having a lower capacity can
reach saturation easily, so that its velocity for the mass transfer zone
is faster than that of the strongly adsorbed species. This difference in
zone velocities causes a so-called chromatographic phenomenon that is
more significant when the travelling distance is longer. If the column
is long enough, for example, at $z = L_2$ as indicated in Figure 10.4, then
there are five distinguishable zones along the column, that is, second
equilibrium-attained zone (zone **I**), second adsorption zone (zone **II**),
first equilibrium-attained zone (zone **III**), first adsorption zone (zone
**IV**) and unsaturated zone (zone **V**). If those zones are expressed from
another viewpoint, then the concentration profiles along the column
are as pictured in Figure 10.5 [13].

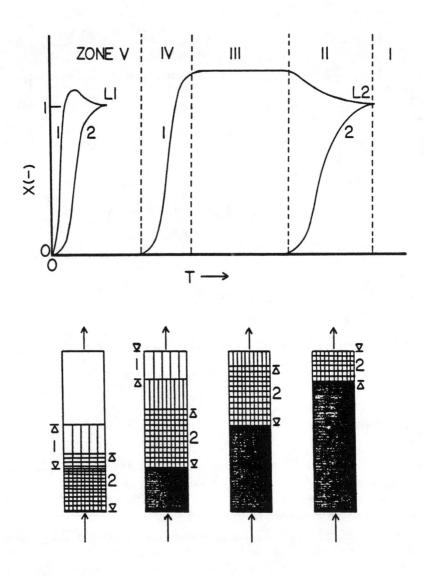

**Figure 10.4.**   Breakthrough Curves and Adsorption Zones for
a Binary System

Obviously, zone **IV** moves faster than zone **II** does, because the first
adsorption zone essentially contains weakly adsorbed species while the
second adsorption zone has both sorbates. The moving velocities of
both zones are not easy to predict. However, a rough estimation can

| Zone | Characteristics |
| --- | --- |
| I | Second Equilibrium-Attained Zone |
| II | Second Adsorption Zone |
| III | First Equilibrium-Attained Zone |
| IV | First Adsorption Zone |
| V | Unsaturated Zone |

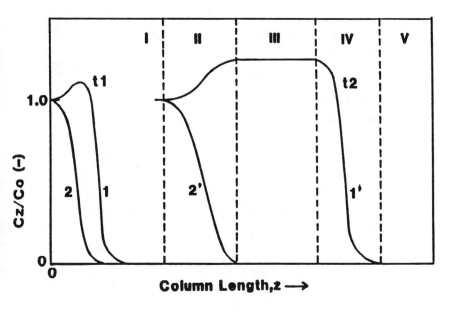

**Figure 10.5.**   Concentration Profiles along the Column for
a Binary System

be achieved according to the equilibrium assumption as follows [14]:

$$V_{mtz} = \left(\frac{\partial z}{\partial t}\right)_C = \frac{V}{1 + \rho_p \dfrac{1 - \epsilon_b}{\epsilon_b}\left(\dfrac{\partial \overline{q}}{\partial C}\right)_C} \qquad (10.14)$$

The value of $\partial \overline{q}/\partial C$ can be calculated approximately by using
$Q_{m1}/C_{m1}$ for the first adsorption zone and $Q_{e2}/C_{02}$ for the second

**Table 10.2.** Experimental Parameters of Fixed Bed Adsorption – Binary System

| Run | $C_0(1)/C_0(2)$ (mmol/l) | MR (–) | FLRT (GPM/ft$^2$) | Temp ($^0$C) | $\epsilon_b$ (–) | $V$ (cm/sec) |
|-----|--------------------------|--------|-------------------|--------------|------------------|--------------|
| E1 | 0.95/0.96 | 0.99 | 5.16 | 23.0 | 0.255 | 1.33 |
| E2 | 0.96/0.97 | 0.99 | 7.23 | 24.0 | 0.246 | 1.94 |
| E3 | 0.98/0.55 | 1.78 | 5.16 | 25.0 | 0.238 | 1.43 |
| E4 | 0.49/0.95 | 0.52 | 5.68 | 24.5 | 0.278 | 1.35 |
| E5 | 0.59/1.46 | 0.40 | 6.04 | 26.0 | 0.264 | 1.51 |
| E6 | 1.04/1.08 | 0.96 | 2.43 | 26.5 | 0.256 | 0.63 |
| F1 | 1.06/1.02 | 1.04 | 6.20 | 24.0 | 0.436 | 0.97 |
| F2 | 0.99/0.57 | 1.74 | 6.97 | 26.5 | 0.447 | 1.06 |
| F3 | 0.51/1.06 | 0.48 | 7.02 | 26.0 | 0.434 | 1.10 |
| F4 | 0.52/1.03 | 0.50 | 8.00 | 26.5 | 0.451 | 1.21 |
| F5 | 1.11/1.14 | 0.97 | 2.32 | 24.0 | 0.413 | 0.38 |
| F6 | 1.98/2.07 | 0.96 | 2.43 | 26.5 | 0.429 | 0.38 |

System: E1 – E6 are Phenol(1)+$p$-Chlorophenol(2)/XAD4; F1 – F6 are Phenol(1)+$p$-Chlorophenol(2)/F400.

adsorption zone. Table 10.2 lists the fixed bed conditions for illustrative examples, and Table 10.3 shows the moving velocities for both MTZ's.

Compared with single component adsorption, the zone velocity of the weakly adsorbed species is significantly faster while that of the strongly adsorbed species is slightly faster in a multicomponent system. This is because the value of $Q_{m1}/C_{m1}$ is smaller than that of $Q_{01}/C_{01}$. This also can explain why the weakly adsorbed species travels forward when the strongly adsorbed species approaches.

Since the calculation of $Q_{m1}/C_{m1}$ requires the value for the maximum excess amount $C_{m1}$, the following equations were developed from the fact that the moving velocities of both sorbates are equal in zone II [13].

$$\frac{\bar{q}_1}{Q_{e1}} = \frac{Q_{e2}/C_{02}}{Q_{e1}/C_{01}}\left(\frac{C_1}{C_{01}} - 1\right) + 1 \qquad (10.15)$$

**Table 10.3.** Analytic Parameters of Fixed Bed Adsorption – Binary System

| Run | $K_{f1}/K_{f2}$ $\times 10^3$ | $D_{s1}/D_{s2}$ $\times 10^8$ | $D_{p1}/D_{p2}$ $\times 10^6$ | Excess (%) | $V_{mtz1}/V_{mtz2}$ $\times 10^3$ |
|-----|------|------|------|-----|------|
| E1 | 4.56/4.28 | 1.95/1.15 | 1.58/1.42 | 35 | 3.50/1.17 |
| E2 | 5.05/4.74 | 1.96/1.18 | 1.62/1.45 | 34 | 4.83/1.63 |
| E3 | 4.42/4.15 | 1.98/0.78 | 1.63/1.46 | 21 | 3.29/0.85 |
| E4 | 5.03/4.72 | 1.62/1.13 | 1.63/1.46 | 53 | 3.14/1.30 |
| E5 | 5.05/4.74 | 1.66/1.64 | 1.67/1.49 | 65 | 3.66/1.74 |
| E6 | 3.70/3.47 | 2.00/1.30 $2.50^*/1.95^*$ | 1.73/1.55 | 35 | 1.71/0.59 |
| F1 | 3.97/3.72 | 2.6/3.1 | 1.57/1.41 | 78 | 1.01/0.51 |
| F2 | 4.34/4.07 | 2.5/2.3 | 1.68/1.50 | 66 | 1.04/0.38 |
| F3 | 4.23/3.97 | 1.7/3.3 | 1.66/1.48 | 161 | 0.86/0.58 |
| F4 | 4.57/4.29 | 1.8/3.2 | 1.68/1.50 | 149 | 1.01/0.66 |
| F5 | 2.73/2.56 | 2.7/3.5 $2.8^*/3.8^*$ | 1.57/1.41 | 84 | 0.39/0.20 |
| F6 | 2.95/2.76 | 4.0/5.2 | 1.68/1.50 | 87 | 0.66/0.35 |

Units in this table are $K_f$: [cm/sec], $D_s$: [cm$^2$/sec], $D_p$: [cm$^2$/sec], $V_{mtz}$: [cm/sec].

\* Estimated from kinetic study by using solid diffusion model.

$$\frac{\overline{q}_2}{Q_{e2}} = \frac{C_2}{C_{02}} \tag{10.16}$$

The maximum excess amount occurs when zone **II** and zone **IV** are separated completely. Therefore, $C_1 \rightarrow C_{m1}$ when $C_2 \rightarrow 0$ under this condition, and Equation (10.17) which is derived from Equation (10.15) can be used to calculate the maximum excess amount of the weakly adsorbed species.

$$\frac{Q_{e2}}{C_{02}}(C_{m1} - C_{01}) + Q_{e1} = K_1 C_{m1}^{1/n_1} \tag{10.17}$$

An iteration method is involved in the procedure of calculation. Results are listed in Table 10.3.

According to the equilibrium model described by Balzli et al.[15], equilibrium between solid and liquid phases is assumed to be established at all points in the bed and the effect of all resistances to mass transfer is neglected. Therefore, the lengths of both adsorption zones are infinitesimal so that two vertical wave fronts are formed by this concept, and the time of their appearance at column length $L$ can be estimated approximately by the following equations:

$$t_{w1} = \frac{W_s L}{V \epsilon_b} \left( \frac{Q_{e2}}{C_{02}} + \frac{Q_{e1}}{C_{m1}} - \frac{Q_{e2} C_{01}}{C_{02} C_{m1}} \right) \qquad (10.18)$$

$$t_{w2} = \frac{W_s L}{V \epsilon_b} \frac{Q_{e2}}{C_{02}} \qquad (10.19)$$

where $W_s$ is the dosage of adsorbent per unit of bed volume.

One example of applying this model is shown in Figure 10.6. Note that the predictions of the equilibrium model approach the experimental breakthrough curves more closely as the column is longer. This is because the mass transfer resistance has a significant effect during the adsorption process. Because of this effect, the solutes are adsorbed more slowly than expected by the equilibrium model, so that sorbate molecules spread over some finite range of adsorption zones. If the column length is short, then the overlap of the adsorption zones can take place. Therefore, competition can exert its influence in the particle by limiting the adsorption of the weakly adsorbed species, and the excess amount is reduced.

The models applied in the following analyses are the combined diffusion model and solid diffusion model, which were developed without the constant pattern assumption and linearized simplification. In the previous chapters, the comparison between these two models showed that the difference in the multicomponent kinetic study is impressive. Therefore, a comparison of these two models was conducted for the binary adsorption in a fixed bed, as illustrated in Figures 10.7 and 10.8. Merk et al.[3] did the same study, except the tortuosity of carbon was set to be 1.0 so that the maximum pore diffusion coefficient was used.

Results show that the difference in prediction by the two models is significant, but not so dramatic as that in the multicomponent kinetic study. Both models produce breakthrough curves at about the same time and predict about the same excess amount, even for the adsorption on the resin bed in which the pore diffusion has a significant contribution to the total intraparticle diffusion. Nevertheless, a flatter increase in the breakthrough curves and a smaller value of the excess

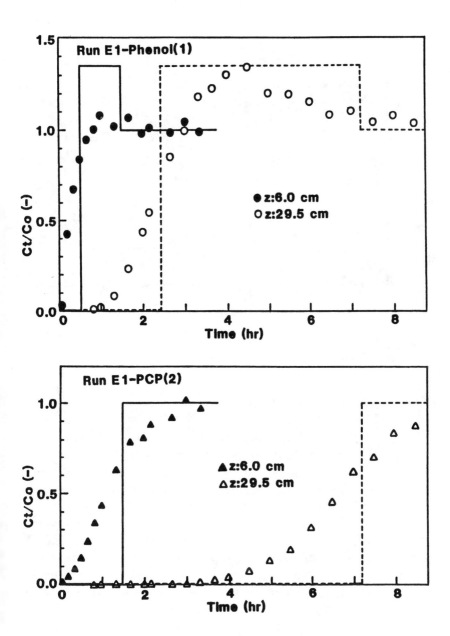

**Figure 10.6.** Prediction of the Equilibrium Model

amount are observed from the simulation of the combined diffusion model.

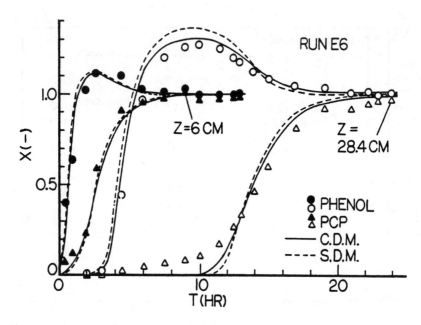

**Figure 10.7.** Comparison between the Solid Diffusion and Combined Diffusion Models

*Adsorption Breakthrough Curves*

Figures 10.9 and 10.10 show the effect of column length on the breakthrough curves from the resin and carbon beds, respectively. The change in column length can be attributed to the moving velocity of the mass transfer zones. Although the mass transfer zone of the weakly adsorbed species (solute **I**) moves faster than that of strongly adsorbed species (solute **II**), these zones overlap each other when the column length is short. Therefore, the adsorption of solute **I** is restricted by the existence of solute **II**, and competition is more important than displacement under this circumstance. When the column is longer, due to the difference in moving velocities, these mass transfer zones are separated gradually. If the column is long enough, the zone of each species can be separated completely. Consequently, some part of the bed is saturated first by adsorbing solute **I** before solute **II** approaches. Under this situation, displacement is the major role of diffusion dynamics, and the pattern of effluent concentration of solute **I** shows an apparent excess curve.

Because the difference between zone velocities is larger in the resin bed than in the carbon bed, the separation of mass transfer zones is

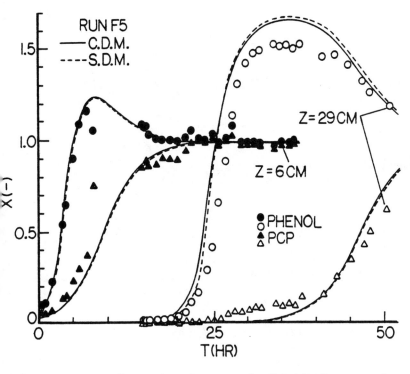

Figure 10.8.    Comparison between the Solid Diffusion and
   Combined Diffusion Models -Example B

more significant in the resin bed at the same column length. Since zone
velocity is a function of interstitial velocity, the effect of this factor
on the breakthrough curves was examined by employing different flow
rates, as displayed in Figures 10.11 and 10.12. Both examples in each
comparison have similar feed concentrations, molar ratio and column
length. Results show that the excess percentage is larger and the
separation of zones is more significant when flow rate is slower. This
is because the contact between sorbate and sorbent is better when the
flow rate is slower; therefore, more sorbate molecules are adsorbed in
the solute-rich part of mass transfer zones  that will reduce adsorption
zones.

   Figures 10.11 and 10.12 show that the excess percentage in the
resin bed is closer to the maximum excess amount than that in the
carbon bed, although the interstitial velocities in the resin bed are
much faster. Part of reason is that the maximum excess percentage
in the resin bed is always less than that in the carbon bed for the
same composition of the mixture. However, this phenomenon can also
be explained by the difference between zone velocities: the larger the

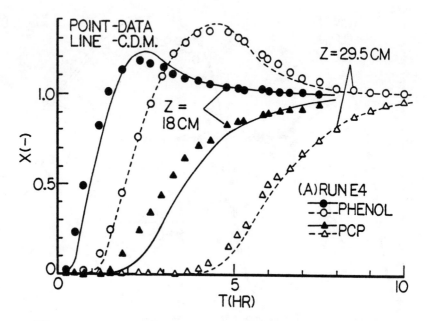

**Figure 10.9.**    Effect of Column Length in the Resin Bed

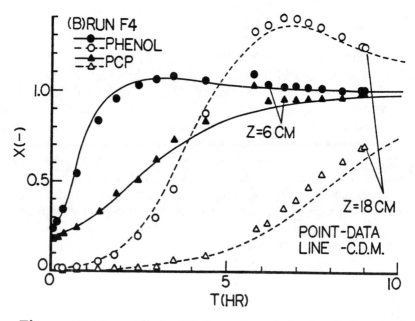

**Figure 10.10.**    Effect of Column Length in the Carbon Bed

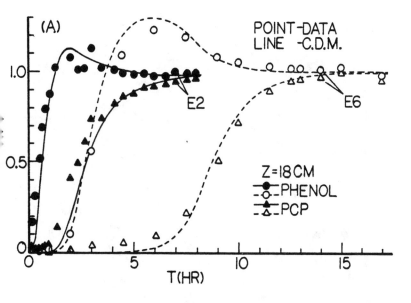

Figure 10.11.    Effect of Flow Rate in the Resin Bed

difference in zone velocities, the more significant the separation of mass transfer zones in a specified column length. If the ratio of zone velocities is 1.0, that is, both solutes in the system have equal zone velocities, then there will be no separation and no displacement along the bed. This extreme condition will happen when both sorbates have the same adsorption behaviors. From Table 10.3, it can be found that the zone velocity of phenol is about three times faster than that of p-chlorophenol in the resin bed. This ratio is only about 2.0 in the carbon bed.

Since the maximum excess percentage is affected by the molar ratio in a mixture, a series of experiments were conducted to consider this case. Although this effect can be read from Table 10.3, a better comparison according to equilibrium model is illustrated in Figure 10.13. Results show that the maximum excess percentages in both systems increase when molar ratios are lower, and that the increase is much faster in the carbon bed than in the resin bed. This is because a larger percentage of the weakly adsorbed species will be displaced by the approach of mass transfer zone containing a higher concentration of the strongly adsorbed species.

Examples in Figure 10.14 show that the mixture having a lower molar ratio appears to have a larger excess percentage at the same col-

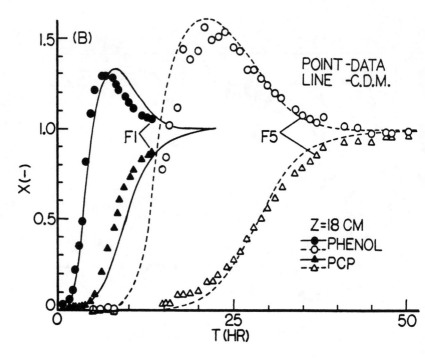

Figure 10.12.    Effect of Flow Rate in the Carbon Bed

umn length when temperatures and interstitial velocities are similar. Figure 10.15 depicts the same tendency. However, due to the larger value of $\partial \bar{q}/\partial C$ results from lower feed concentration (from Equation (10.14)), the ratio of zone velocities decreases when molar ratio decreases, as listed in Table 10.3. The separation of adsorption zones is less significant for lower molar ratio at a specified column length. This is because the mixture containing lower molar ratio has larger maximum excess percentage and lower ratio of zone velocities, and a longer column will be necessary to achieve a complete separation of mass transfer zones.

If both sorbates in the mixture have higher feed concentrations when molar ratio is kept constant, then the value of maximum excess percentage increases, too. However, this increase is significant when the molar ratio is lower. The reason for this increase is attributed to the mutual effect from the nonlinear characteristics of isotherms. When the molar ratio is lower, the effect of nonlinear isotherm from the strongly adsorbed species increases so that the maximum excess percentage increases significantly. If the condition of equal molar feed is considered, then the effect of increasing concentrations can be ob-

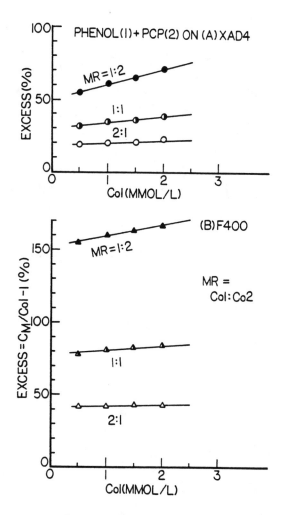

**Figure 10.13.**    Maximum Excess Amount for Different
Compositions of Mixture

served in Figure 10.16.

Merk *et al.* [3] in their binary adsorption studies found that if the molar ratio was kept about 1.0, then the excess amount decreased at the same column length when both concentrations in the mixture increased. They concluded that this phenomenon happens because the internal mass transfer resistance increases significantly with increasing influent concentration. The same result is shown in Figure 10.16 in which the experimental examples have the same column length, molar ratio and interstitial velocity, except that one mixture has double the feed concentrations of the other. From Table 10.3, however, the

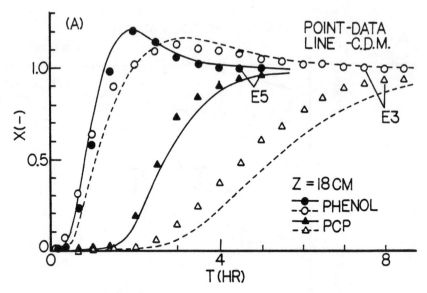

**Figure 10.14.**    Molar Ratio Effect in the Resin Bed

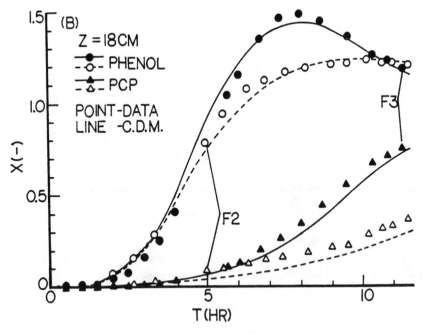

**Figure 10.15.**    Molar Ratio Effect in the Carbon bed

reasons for the lower excess percentage for the higher feed concentrations seem to be the lower ratio of zone velocities and the faster

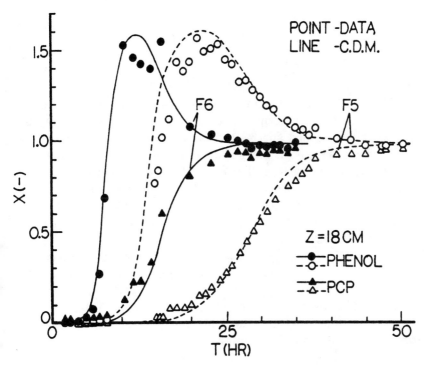

**Figure 10.16.**    Input Concentration Effect in the Carbon Bed

moving velocities of both adsorption zones. Under this circumstance, the competition effect will be enhanced because of more overlap between adsorption zones. Therefore, the separation of zones is less significant, the raise of breakthrough curves is earlier and the excess percentage is smaller for Run F5 than for Run F6, at the same column length.

*Empirical Model*

To estimate the breaktime of each species in a binary system, the equations derived from single component systems were extended as follows.

$$\left(\frac{dX_i}{dt}\right)_{X_i=0.5} = \left(\frac{\overline{D}_{i,int}k_{f_i}\overline{a}_p\epsilon_b}{R_p^2 P_{bi}(1-\epsilon_b)}\right)^{1/2} \tag{10.21}$$

$$t_{bi} = t_{wi} - \frac{1}{2}(0.8T_{(mtz)_i}) = \frac{L}{V_{(mtz)_i}} - \frac{0.4}{(dX_i/dt)_{X_i=0.5}} \tag{10.22}$$

where

$$P_{bi} = \frac{q_i\rho_p(1-\epsilon_b)}{C_i\epsilon_b} \tag{10.23}$$

$$\overline{D}_{i,int} = D_{si} + D_{pi}\frac{C_i\epsilon_p}{q_i\rho_p} \tag{10.24}$$

$$T_{(mtz)_i} = \frac{1}{(dX_i/dt)_{X_i=0.5}} \tag{10.25}$$

In a binary system, the solute partition parameters are reduced by the competition and replacement effects during the process, especially for the weakly adsorbed species (solute I). Consequently, the breakthrough curve of solute I is much sharper in the binary system than in the single component system. The concentration of solute I in the first adsorption zone is from 0 to $C_{m1}$, and that of solute II in the second adsorption zone is from 0 to $C_{02}$. Therefore, $C_1 = C_{m1}$, $q_1 = Q_{m1}$ for solute I and $C_2 = C_{02}$, $q_2 = Q_{e2}$ for solute II were used in Equations (10.24) and (10.25).

For a complete procedure, however, more steps are necessary to depict the excess part of curve I. From the previous analyses, the excess amount was found to be affected by the overlapping extent of both adsorption zones, which is determined by column length, zone velocities and zone lengths. The following equation including these factors was derived to judge overlap.

$$\frac{T_{mtz1} + T_{mtz2}}{2} \begin{cases} > t_{w2} - t_{w1} & \text{overlap} \\ \leq t_{w2} - t_{w1} & \text{no overlap} \end{cases} \tag{10.26}$$

If the column is not long enough so that the overlap occurs, the excess amount of solute I can not reach its maximum value, $X_{m1}$, due to competition. However, on the excess curve there is an upper value, $X_{u1}$, which can be estimated as follows:

$$X_{u1} = \frac{t_{w2} - t_{w1} + \dfrac{T_{mtz1}}{2} + \dfrac{T_{mtz2}}{2}\left(\dfrac{X_{m1} + 1.0}{X_{m1} - 1.0}\right)}{\dfrac{T_{mtz1}}{X_{m1}} + \dfrac{T_{mtz2}}{X_{m1} - 1.0}} \tag{10.27}$$

where $X_{u1} = C_{u1}/C_{01}$, $X_{m1} = C_{m1}/C_{01}$, and $1.0 \leq X_{u1} \leq X_{m1}$. An example applying this method is illustrated in Figure 10.17. A comparison between observed and calculated data is shown in Table 10.4. Again, the deviation of predicted values is lower when the column is longer, because the relative error is smaller.

*CONCLUSIONS*

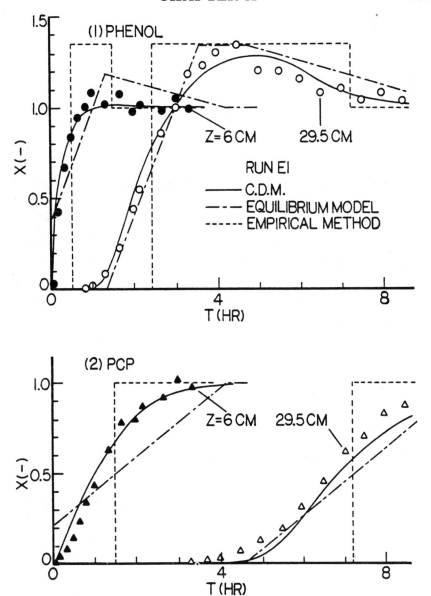

**Figure 10.17.**    The Application of the Empirical Model

Adsorption behaviors of a mixture containing two phenolic compounds adsorbed onto two different types of adsorbent, XAD4 and F400, were investigated in this study. Because of the selectivity of the sorbents, the sorbates in a mixture can be separated gradually along the column. Usually, the larger difference between zone velocities, the slower zone velocities and the smaller maximum excess percentage

**Table 10.4.**    Prediction of Breaktime in Binary Systems

| Run | $Y_{e1}/Y_{e2}$ (-) | $Sp_1/Sp_2^*$ $\times 10^4 (sec^{-1})$ | $T_{z1}/T_{z2}^{**}$ (min) | $L$ (cm) | $T_{b1}/T_{b2}$ (min) | $T_{b1}/T_{b2}^\dagger$ (min) |
|---|---|---|---|---|---|---|
| E1 | 0.281/0.968 | 1.39/0.562 | 120/296 | 29.5 | 96/313 | 85/281 |
|    |             |            |         | 18.0 | 40/145 | 37/138 |
| E2 | 0.280/0.968 | 1.41/0.576 | 118/289 | 29.5 | 57/194 | 50/162 |
|    |             |            |         | 18.0 | 16/ 73 | 10/ 68 |
| E3 | 0.335/0.951 | 1.24/0.364 | 134/458 | 29.5 | 99/414 | 80/365 |
|    |             |            |         | 18.0 | 39/181 | 28/148 |
| E4 | 0.212/0.984 | 1.30/0.640 | 128/260 | 29.5 | 109/285 | 74/274 |
|    |             |            |         | 18.0 | 47/133 | 25/115 |
| E5 | 0.200/0.986 | 1.34/0.819 | 124/203 | 26.0 | 71/175 | 64/169 |
|    |             |            |         | 18.0 | 34/ 96 | 30/ 97 |
| E6 | 0.280/0.968 | 1.31/0.561 | 127/297 | 28.4 | 233/705 | 213/635 |
|    |             |            |         | 18.0 | 129/403 | 120/315 |

From Longest Column: $S1 = 23.7\%$, $S2 = 11.9\%$.
At Column Length $L = 18.0$ cm: $S1 = 42.1\%$, $S2 = 16.4\%$.

| Run | $Y_{e1}/Y_{e2}$ (-) | $Sp_1/Sp_2^*$ $\times 10^4 (sec^{-1})$ | $T_{z1}/T_{z2}^{**}$ (min) | $L$ (cm) | $T_{b1}/T_{b2}$ (min) | $T_{b1}/T_{b2}^\dagger$ (min) |
|---|---|---|---|---|---|---|
| F1 | 0.291/0.812 | 0.284/0.235 | 587/710 | 24.0 | 251/493 | 243/532 |
|    |             |            |         | 18.0 | 130/298 | 113/315 |
| F2 | 0.387/0.731 | 0.269/0.181 | 618/919 | 25.2 | 309/672 | 305/- - - |
|    |             |            |         | 18.0 | 150/376 | 135/580 |
| F3 | 0.170/0.904 | 0.207/0.239 | 804/698 | 24.1 | 240/466 | 280/595 |
|    |             |            |         | 18.0 | 98/277 | 155/360 |
| F4 | 0.178/0.898 | 0.234/0.256 | 713/650 | 25.3 | 220/421 | 232/480 |
|    |             |            |         | 18.0 | 75/224 | 125/280 |
| F5 | 0.280/0.821 | 0.228/0.198 | 732/840 | 29.0 | 1242/2064 | 1340/1970 |
|    |             |            |         | 18.0 | 660/1153 | 561/1050 |
| F6 | 0.287/0.816 | 0.390/0.342 | 427/487 | 28.5 | 697/1137 | 667/1048 |
|    |             |            |         | 18.0 | 377/646 | 370/570 |

From Longest Column: $S1 = 7.3\%$, $S2 = 12.4\%$.
At Column Length $L = 18.0$ cm: $S1 = 24.5\%$, $S2 = 20.3\%$.

---

\* Sp = Slope of curve at $X_i = 0.5$ was calculated by using Equation (10.21).

\*\* $T_z$ is the travelling time of mass transfer zone passing a fixed point in the column.

† Estimated from experimental data.

will assure a complete separation of mass transfer zones in a shorter column. Those factors are related to the equilibrium isotherms, inlet concentrations, flow rate and molar ratio. Therefore, different operating conditions were considered in this study. Table 10.5 describes the effects of those operative conditions in the fixed bed adsorber.

While the equilibrium isotherms determine the breakthrough time, the mass transfer resistances affect the shape of the curves. If the effect of rate parameters is considered, then surface diffusion is more important than pore diffusion and external film mass transfer. Because surface diffusion flux was found to be about 80% of the intraparticle diffusion flux for resin adsorption and 95% for carbon adsorption, and since intraparticle diffusion is the rate controlling factor in the adsorption zones, the estimation of surface diffusion coefficients is important for a precise modeling simulation. The relation of surface diffusion coefficient and inlet liquid concentration derived from kinetic study was shown to be applicable to binary adsorption in fixed beds. Agreements between modeling predictions and experiments are fairly good in general, suggesting the validity of the theory and the applicability of the combined diffusion model.

**Table 10.5.** Effects of Operative Conditions in Fixed Bed Adsorber Binary System

| Factor | Chromatographic phenomenon | Reason |
|---|---|---|
| Adsorbate | The curve **I** shows a steeper pattern and appears earlier than the curve **II** does. | The solute **I**, a weakly adsorbed species, has lower adsorptive capacity, so that its $L_{mtz}$ is shorter and $V_{mtz}$ is faster than that of the solute **II**, a strongly adsorbed species. |
| | The curve **I** exhibits an overshooting pattern while the curve **II** rises gradually. | The excess part of solute **I** is the amount displaced by solute **II**. |
| Column length ($L$) | Both curves are steeper when $L$ is shorter. | The constant pattern is not formed yet. |
| | The chromatographic effect is more significant when $L$ is longer. | The separation of zones is more apparent when the traveling distance is longer, if $V_{mtz}(1) \neq V_{mtz}(2)$. |
| | The overshooting pattern on the curve **I** is less apparent when $L$ is shorter. | Adsorption zones overlap each other so that competition is more significant than displacement. |
| | The maximum point of the curve **I** approaches the maximum excess percentage when $L$ is longer. | Adsorption zones separate from each other because $V_{mtz}(1) > V_{mtz}(2)$; displacement is more significant than competition. |
| Interstitial velocity | Both curves are steeper when $V$ is higher. | Both $L_{mtz}$'s are shorter because $K_f = f(V)$. |

(*continued*)

**Table 10.5.**   Effects of Operative Conditions in Fixed Bed Adsorber –
Binary System (*continued*)

| Factor | Chromatographic phenomenon | Reason |
|---|---|---|
| $(V)$ | The chromatographic effect is less significant when $V$ is higher. | Both $V_{mtz}$'s are faster because $V_{mtz} = f(V)$. |
| | The excess percentage on the curve **I** is lower when $V$ is higher. | The adsorption zones overlap each other and a longer traveling distance is needed for separation, because $V_{mtz} = f(V)$. |
| Molar ratio (MR) | The excess percentage on the curve **I** is lower when MR is higher. | The maximum excess percentage of solute **I** is lower when MR is higher. |
| | The chromatographic effect is more significant when MR is higher. | The difference between both $V_{mtz}$'s is larger when MR is higher. |
| | Both curves are flatter when MR is higher. | The interference between both zones is less significant due to the larger difference between $V_{mtz}$'s. |
| Concentration $(C_0)$ | Both curves are steeper when both $C_0$'s are higher. | Both $D_s$'s are larger because $D_s = f(C_0)$, so that both $L_{mtz}$'s are shorter. |
| | The chromatographic effect is less significant if both $C_0$'s are higher. | Both $P_b$'s are smaller so that both $V_{mtz}$'s are faster. |
| | The excess percentage on the curve **I** is lower when both $C_0$'s are higher. | Both $V_{mtz}$'s are faster when $C_0$'s are higher, so that zones are closer to each other. |

(*continued*)

**Table 10.5.**  Effects of Operative Conditions in Fixed Bed Adsorber Binary System (*continued*)

| Factor | Chromatographic phenomenon | Reason |
|---|---|---|
| Adsorption system (Sorbates -Sorbent) | Both curves appear earlier and the chromatographic effect is less significant when the adsorbent has lower adsorption capacity. | Both $V_{mtz}$'s are faster due to smaller $P_b$'s, because $V_{mtz} = f(1/P_b)$. |
| | Both curves are steeper if the adsorbent has lower adsorption capacity. | Both $L_{mtz}$'s are shorter due to smaller $P_b$'s and/or larger $D_s$'s. |
| | The excess percentage on the curve **I** is lower when the adsorbent has lower adsorption capacity. | The maximum excess percentage is lower due to the weaker adsorptive characteristics of the adsorbent. |

*REFERENCES*

1. Crittenden, J.C., and W.J. Weber, Jr., *ASCE. EE.*, **104**, 1175 (1978).
2. Crittenden, J.C., B.W.C. Wong, W.E. Thacker, V.L. Snoeyink, and R.L. Hinrichs, *J. WPCF.*, **52**, 2780 (1980).
3. Merk, W., W. Fritz, and E.U. Schlünder, *Chem. Eng. Sci.*, **36**, 743 (1981).
4. Liapis, A.I., and D.W.T. Rippin, *AIChE. J.*, **25**, 455 (1979).
5. Mansour, A., D.U. von Rosenberg, and N.D. Sylvester, *AIChE. J.*, **28**, 765 (1982).
6. Hsieh, J.S.C., R.M. Turian, and C. Tien, *AIChE. J.*, **23**, 263 (1977).
7. Miura, K., H. Kurahashi, Y. Inokuchi, and K. Hashimoto, *J. Chem. Eng. Japan*, **12**, 281 (1979).
8. Moon, H., and W.K. Lee, *Chem. Eng. Sci.*, **41**, 1995 (1986).
9. Barba, D., G. Del Re, and P.U. Foscolo, *Chem. Eng. J.*, **26**, 33 (1983).
10. Radke, C.J., and J.M. Prausnitz, *AIChE. J.*, **18**, 761 (1972).
11. Hindmarsh, A.C. *GEAR: Ordinary Differential Equation System*

*Solver, UCID-30001 Rev. 3.* Lawrence Livermore Laboratory, P.O.Box 808, Livermore, CA (1974).

12. Crittenden, J.C. *Mathematic Modeling of Fixed Bed Adsorber Dynamics – Single Component and Multicomponent.* Ph.D. Dissertation, University of Michigan, Ann Arbor, MI (1976).

13. Cooney, D.O., and F.P. Strusi, *Ind. Eng. Chem. Fundam.*, **11**, 123 (1972).

14. Coulson, J.M., and J.F. Richardson, *Chemical Engineering.* 2nd ed., Pergamon Press, Oxford, New York (1979).

15. Balzli, M.W., A.I. Liapis, and D.W.T. Rippin, *Trans. I. Chem. E.*, **56**, 145 (1978).

# CHAPTER XI

## COMPARISON OF AIR AND WATER ADSORPTION SYSTEMS

Contributing Author: B.G. Pierce

## *INTRODUCTION*

Adsorption has been handled separately in this book for air and water adsorption systems. This is because there is a general division of contaminants into those that occur in air and those that occur in water. However, it is possible to provide data to allow a direct comparison of air and water adsorption systems if the same adsorbate is utilized for experimental studies in both media. The nature of benzene renders it a pollutant that occurs in both air and water, and thus allows one to compare isotherm and diffusion data collected from both media. Waterborne benzene is usually associated with water contamination from underground storage tanks. Benzene, a known carcinogen and contamination of potable water sources, is not desirable. The occurrence of benzene in air is usually associated with volatilization. Carbon adsorption can be used to reduce the concentration to allowable levels in both air and water.

The vapor pressures at which benzene volatilizes appear in Table 11.1. It should be noted that the vapor pressures associated with what are generally considered "ambient" temperatures are high, though not as high as other organic air contaminants, such as carbon tetrachloride, toluene and *p*-xylene. This characteristic of benzene is the basis for the fact that benzene is a cross-over contaminant; the volatility of benzene renders it a contaminant in air while the vapor pressure it exerts is low enough to maintain it as a contaminant in water. The evaluation of the sorptive behavior of benzene was accomplished by utilization of a long column apparatus, in air (Chapter IX) and water (this chapter). The long columns were utilized for both isothermal and kinetic data evaluations.

## *METHOD AND RESULTS*

### *Adsorption of Benzene in Liquid-Solid System*

Table 11.1.   Vapor Pressure of Benzene at Different Temperatures [5]

| Temperature ($^0$F) | Vapor Pressure (lbs/in$^2$) |
|---|---|
| 50 | 0.880 |
| 60 | 1.170 |
| 70 | 1.534 |
| 80 | 1.987 |
| 90 | 2.544 |
| 100 | 3.224 |
| 110 | 4.045 |
| 120 | 5.028 |

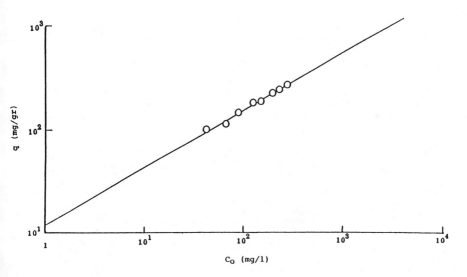

Figure 11.1.   Isotherms for Benzene/XAD-4 and Benzene/F400

Isotherm. Figure 11.1 depicts the adsorption isotherms for benzene on resin XAD-4 and carbon F400, respectively. The parameters of the Freundlich model indicate that the carbon F400 has a higher capacity for benzene than does the resin XAD-4. This result is expected, as the surface area for carbon (1100 m$^2$/g) was greater than the surface area for the resin (750 m$^2$/g).

Mass Transfer Zones.   The mass transfer zone lengths appear in

Tables 11.2 and 11.3. The values were derived utilizing the method of Hand *et al.* [2]. The length of the mass transfer zone varies for the different runs, due to variation in the parameters used for calculation (such as bed void fraction, Stanton number and Biot number). As can be seen from examination of Tables 11.2 and 11.3, the lengths of the mass transfer zones for carbon are generally greater than those for resin.

**Table 11.2.**    Mass Transfer Zone Lengths – XAD-4 Resin

| Run Number | MTZ Length (cm) | % Fully Developed |
|:---:|:---:|:---:|
| R1 | 9.88 | 100 |
| R2 | 8.74 | 100 |
| R3 | 9.54 | 100 |
| R4 | 8.61 | 100 |
| R5 | 9.47 | 100 |
| R6 | 10.05 | 100 |
| R7 | 8.84 | 100 |
| R8 | 9.53 | 100 |

**Table 11.3.**    Mass Transfer Zone Lengths – F400 Carbon

| Run Number | MTZ Length (cm) | % Fully Developed |
|:---:|:---:|:---:|
| C1 | 16.67 | 94.28 |
| C2 | 16.99 | 93.33 |
| C3 | 16.44 | 93.35 |
| C4 | 18.18 | 92.00 |
| C5 | 17.78 | 92.90 |
| C6 | 19.26 | 91.80 |
| C7 | 15.85 | 95.02 |
| C8 | 15.62 | 95.10 |
| C9 | 14.23 | 96.77 |

Kinetics. The isotherms data were used along with the film mass transfer coefficients in the MADAM computer model [2] to produce breakthrough curves that could be compared to those determined experimentally. The reason for this was to determine the remaining parameter used in the model, the surface diffusion coefficient, using a curve-fitting technique, so that the value of the coefficient is the best fit to the experimental breakthrough curves. The results obtained are presented in Table 11.4 for XAD-4 resin and Table 11.5 for F400 carbon. The deviations have a range of 2.78% to 17.68% for the XAD-4, and 2.88% to 10.10% for the F400. The $D_e$ values range from $9.0 \times 10^{-9}$ cm$^2$/sec to $3.2 \times 10^{-8}$ cm$^2$/sec for XAD-4 resin and $2.5 \times 10^{-8}$ cm$^2$/sec to $1 \times 10^{-7}$ cm$^2$/sec for F400 carbon.

**Table 11.4.**   Standard Deviations of Predicted Breakthrough Curves – XAD-4 Resin

| Run# | $C_0$ (mg/l) | $D_e \times 10^8$ (cm$^2$/sec) | $k_f \times 10^3$ (cm/sec) | Bed Depth (cm) | Standard Deviation (%) |
|------|------|------|------|------|------|
| R1 | 42 | 0.9 | 5.63 | 6.0 | 6.92 |
|    |    |     |      | 12.0 | 17.68 |
| R2 | 202 | 2.9 | 5.72 | 6.0 | 3.79 |
|    |    |     |      | 12.0 | 4.59 |
| R3 | 233 | 3.1 | 6.13 | 6.0 | 3.95 |
|    |    |     |      | 12.0 | 3.31 |
| R4 | 90 | 1.5 | 5.51 | 6.0 | 4.08 |
|    |    |     |      | 12.0 | 5.06 |
| R5 | 125 | 2.1 | 5.56 | 6.0 | 7.62 |
|    |    |     |      | 12.0 | 5.70 |
| R6 | 153 | 2.2 | 6.21 | 6.0 | 4.84 |
|    |    |     |      | 12.0 | 2.78 |
| R7 | 67 | 1.4 | 5.73 | 6.0 | 6.95 |
|    |    |     |      | 12.0 | 7.13 |
| R8 | 268 | 3.2 | 5.79 | 6.0 | 8.44 |
|    |    |     |      | 12.0 | 5.67 |

The MADAM model used to determine the value for the diffusion

**Table 11.5.**   Standard Deviations of Predicted Breakthrough Curves
– F400 Carbon

| Run# | $C_0$ (mg/l) | $D_e \times 10^8$ (cm$^2$/sec) | $k_f \times 10^3$ (cm/sec) | Bed Depth (cm) | Standard Deviation (%) |
|------|------|------|------|------|------|
| C1 | 77 | 3.0 | 4.01 | 6.0 | 4.58 |
|    |    |     |      | 12.0 | 3.45 |
| C2 | 130 | 5.0 | 4.06 | 6.0 | 6.37 |
|    |    |     |      | 12.0 | 2.88 |
| C3 | 251 | 10.0 | 4.21 | 6.0 | 6.99 |
|    |    |     |      | 12.0 | 4.05 |
| C4 | 194 | 6.5 | 3.82 | 6.0 | 3.99 |
|    |    |     |      | 12.0 | 10.09 |
| C5 | 110 | 4.0 | 3.65 | 6.0 | 4.60 |
|    |    |     |      | 12.0 | 9.68 |
| C6 | 63 | 2.5 | 3.68 | 6.0 | 4.45 |
|    |    |     |      | 12.0 | 6.49 |
| C7 | 216 | 9.2 | 3.93 | 6.0 | 5.16 |
|    |    |     |      | 12.0 | 3.38 |
| C8 | 100 | 5.0 | 3.94 | 6.0 | 7.07 |
|    |    |     |      | 12.0 | 8.23 |
| C9 | 148 | 6.0 | 3.62 | 6.0 | 7.86 |
|    |    |     |      | 12.0 | 4.27 |

coefficients is a surface diffusion model and assumes a constant internal diffusion coefficient value. The model is used in an orthogonal collocation computer program, to define the adsorption process and determine the concentrations along the length of the simulated column at discrete points defined by the program. By solution of a series of differential and polynomial equations defining boundary conditions and interior concentration gradients incorporated with the surface diffusion model, the desired concentrations are calculated both on the interior of the particle (the discrete points) over the radius of the particle and externally over time, the external concentrations at desired times and specific points along the column being reported [3].

*Comparison of Benzene Adsorption in Water and Air Systems*

Sarlis [4] used benzene vapor, as well as other organic air contaminants, in the examination of adsorption phenomena by gravimetric analysis (Chapter IV). Isothermal parameters, as well as kinetic values were obtained for benzene on XAD-4 resin and beaded activated carbon (BAC). Comparisons of the isothermal parameters appear in Table 11.6 with graphical representations appearing in Figure 11.2.

**Table 11.6.**   Isothermal Parameters Comparison for Benzene

| System | Medium | $K$ | $1/n$ | $r^2$ |
|---|---|---|---|---|
| Reported: | | | | |
| Benzene/BAC | Air | 62.01 | 0.190 | 0.98 |
| Benzene/XAD-4 | Air | 2.06 | 0.469 | 0.93 |
| Normalized to Water Units: | | | | |
| (for Isothermal Comparison) | | | | |
| Benzene/BAC | Air | 184.81 | 0.190 | 0.98 |
| Benzene/F400 | Water | 28.2 | 0.483 | 0.92 |
| Benzene/XAD-4 | Air | 29.78 | 0.469 | 0.93 |
| Benzene/XAD-4 | Water | 12.1 | 0.549 | 0.98 |

The isotherms can be seen to have some qualities in common between the two media. First, for both air and water, the carbon isotherms have higher intercepts (and correspondingly high capacities) than their resin counterparts. This was to be expected based on the surface area comparison made previously. Secondly, comparison of the relative slopes of the isotherms for the adsorbents demonstrates that XAD-4 resin has a constantly higher slope and a correspondingly higher isotherm exponent value than carbon, even though the types of carbon used as adsorbent in each study were different.

*Breakthrough Curve Comparison*

Breakthrough curves were generated by computer simulation in both air and water for comparison purposes. The vapor phase concentrations in air were normalized to equal units in water in order that

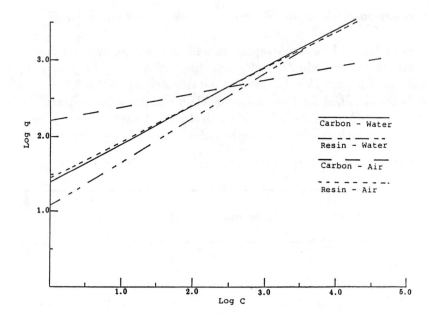

**Figure 11.2.** Isotherms of Benzene in the Air and Water Adsorption Systems

further comparisons could be made. These values appear in Table 11.7. The method used to generate the conversion was:

**Table 11.7.** Normalized Solute Concentrations in Air

| Concentration in Air (ppm) | Concentration in Water (mg/l) |
| --- | --- |
| 85 | 0.276 |
| 352 | 1.123 |
| 425 | 1.379 |
| 850 | 2.758 |
| 2102 | 6.821 |
| 4236 | 13.745 |
| 8437 | 27.380 |

$$C_g = C_l \times \frac{1 \times 10^{-6} \text{ lit. Benz.}}{1 \text{ lit. Air}} \times \frac{1 \text{ lit. Water}}{1 \text{ mg Benz.}} \times$$
$$\frac{1 \text{ mole Benz.}}{22.4(298/273) \text{ lit.Benz.}} \times \frac{78000 \text{ mg Benz.}}{1 \text{ mole Benz.}} \quad (11.1)$$

where $C_g$ = concentration in gas phase (ppm), and $C_l$ = concentration in liquid phase (mg/l). Tabular presentation of the standard physical parameters used appears in Table 11.8.

Table 11.8.  Standard Parameters Used in Breakthrough Curve Analysis in Air and Water Phases

| Parameter | Gas-Solid | Liquid-Solid |
|---|---|---|
| $C_0$ | 6160 ppm | 19.7 mg/l |
| $K$ | 2.06 | 12.1 |
| $1/n$ | 0.4695 | 0.5488 |
| $k_f$ (cm/sec) | 3.0 | $3.0 \times 10^{-3}$ |
| Flow rate | 3800 cm$^3$/min | 100 ml/min |
| $D_e$ (cm$^2$/sec) | $2.2 \times 10^{-7}$ | $6.5 \times 10^{-9}$ |

Figures 11.3 and 11.4 depict the breakthrough curve comparisons for XAD-4 resin. Figure 11.3 uses velocities found in typical air adsorption and Figure 11.4 uses velocities typical in water adsorption. The data for the air breakthrough curves were taken from Yeh [5] using his effective model (Chapter IX), and the data for the water breakthrough curves was generated on the MADAM computer model as previously described.

As can be seen from the two figures, the air breakthrough curves are much sharper than are those for water. This is due to the much higher diffusivities for benzene in the air adsorption system, compared to the water adsorption system. Also the air breakthrough curves are displaced to the right. This is due to the difference in the isotherm data, the $K$ in the air being much greater than in water and the exponent being less in air than in water, indicating an increased capacity for the adsorbent in the air/benzene system.

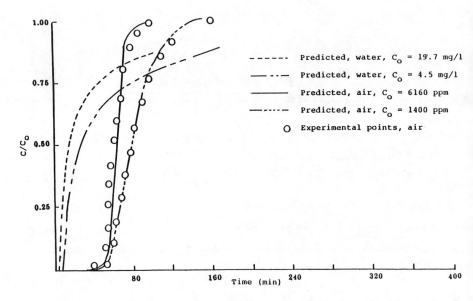

**Figure 11.3.**    Predicted Breakthrough Curves for Air to Water Comparison ($F = 3800$ cm$^3$/min)

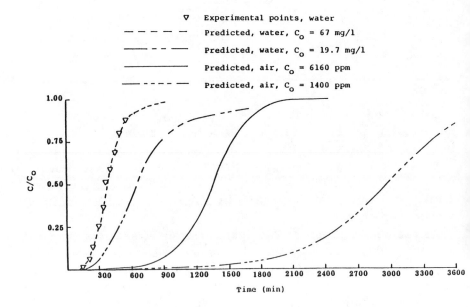

**Figure 11.4.**    Predicted Breakthrough Curves for Air to Water Comparison ($F = 100$ cm$^3$/min)

An analysis of the parameters involved in breakthrough curve determination; i.e., the diffusion parameter $De$, isothermal parameter $1/n$, and kinetic parameters, was performed to determine the relative effect of each parameter on the breakthrough curve. The results of the analyses appear in Tables 11.9 through 11.11, with graphical demonstration of the effects appearing in Figures 11.5 through 11.10.

Table 11.9.  Percent Change in Breakthrough Due to Velocity in Air and Water Phases

| Phase | Velocity | $C/C_0 = 0.25$ | $C/C_0 = 0.50$ | $C/C_0 = 0.75$ |
|---|---|---|---|---|
| Water | 100 ml/min | (970)* | (1110) | (1350) |
|  | 3800 ml/min | -99.2% | -98.6% | -96.8% |
| Air | 100 cm$^3$/min | +3359.7% | +3802.0% | +4101.1% |
|  | 3800 cm$^3$/min | (83.75) | (86.75) | (92.00) |

* Standard Parameter Reference in Parentheses (min).

Table 11.10.  Percent Change in Breakthrough Due to Isothermal Parameter $1/n$ in Air and Water Phases

| Phase | $1/n$ | $C/C_0 = 0.25$ | $C/C_0 = 0.50$ | $C/C_0 = 0.75$ |
|---|---|---|---|---|
| Water | 0.400 | -33.9% | -37.9% | -40.4% |
|  | 0.5488 | (805)* | (950) | (1135) |
|  | 0.6000 | +15.5% | +16.8% | +21.1% |
| Air | 0.3571 | -63.28% | -62.25% | -60.33% |
|  | 0.4000 | -48.66% | -48.13% | -46.20% |
|  | 0.4695 | (83.7) | (86.7) | (92.0) |

* Standard Parameter Reference in Parentheses (min).

Table 11.11.  Percent Change in Breakthrough Due to Diffusivity Parameter $D_e$ in Air and Water Phases

| Phase | $D_e(\pm\%)$ | $C/C_0 = 0.25$ | $C/C_0 = 0.50$ | $C/C_0 = 0.75$ |
|-------|--------------|----------------|----------------|----------------|
| Water | -20% | -6.6% | -2.4% | +1.2% |
|       | 0%   | (450)* | (625) | (840) |
|       | +20% | +2.2%+0.8% | -1.2% | |
| Air   | -20% | -3.2% | -2.9% | +2.6% |
|       | 0%   | (30.5) | (33.5) | (38.5) |
|       | +20% | +3.2% | +1.5% | -2.6% |

\* Standard Parameter Reference in Parentheses (min).

As can be seen from the data, the velocity of the medium, air or water, had the greatest effect, followed by the isothermal parameter $1/n$, with the diffusion parameter $D_e$ generating the least effect. This relationship holds true for both media. An increase in velocity decreases the time to breakthrough and decreases the overall breakthrough curve, while an increase in the thermodynamic parameter $1/n$, increases the time to breakthrough and increases the overall breakthrough curve. An increase in the diffusion parameter $D_e$ decreases the overall breakthrough curve; however, there is no significant effect on the time of breakthrough with a variance in $D_e$.

A comparison of the mass transfer zones can also be made. It is quite apparent that the mass transfer zone lengths associated with the air adsorber columns will be much shorter than those associated with water adsorption columns. In terms of design of adsorber systems, this has a large impact on the length of the column necessary to achieve a 100% breakthrough curve profile, the air column achieving this more rapidly than the water column.

*Comparison of the Intraparticle Diffusivity in Both Systems*

Data collected on the diffusion coefficient for different air and water systems and reported in Chapters IV and V can be compared if a dimensionless diffusion coefficient can be defined that accounts for differences in the physical properties of the two media. To develop

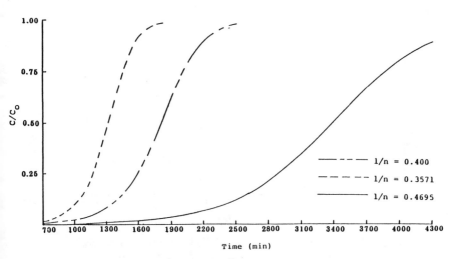

**Figure 11.5.** Effect of Isothermal Parameter $1/n$ Variance on Breakthrough Curves for Air at Water Flow Rate ($F = 100$ cm$^3$/min)

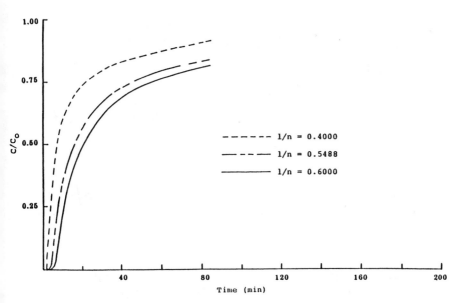

**Figure 11.6.** Effect of Isothermal Parameter $1/n$ Variance on Breakthrough Curves for Water at Air Flow Rate ($F = 3800$ cm$^3$/min)

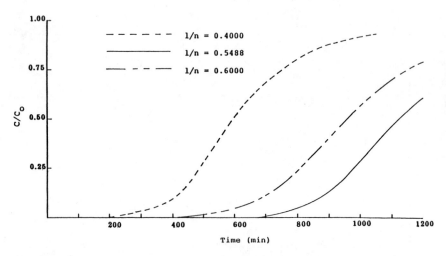

**Figure 11.7.**    Effect of Isothermal Parameter $1/n$ Variance on Breakthrough Curves for Water at Water Flow Rate ($F = 100$ cm$^3$/min)

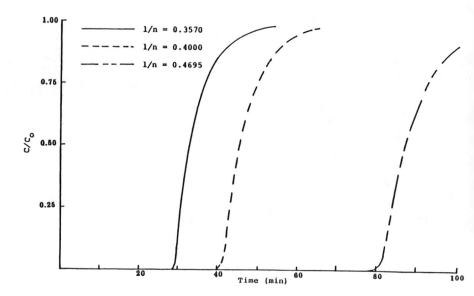

**Figure 11.8.**    Effect of Isothermal Parameter $1/n$ Variance on Breakthrough Curves for Air at Air Flow Rate ($F = 3800$ cm$^3$/min)

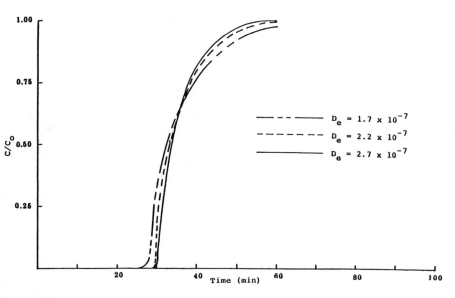

**Figure 11.9.**    Effect of Diffusion Parameter $D_e$ Variance on Breakthrough Curves in the Air System

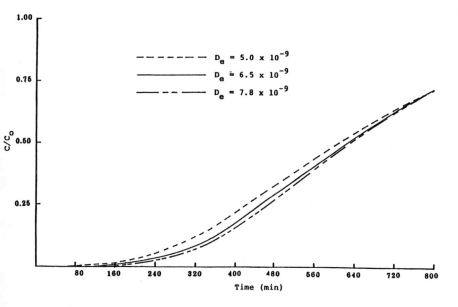

**Figure 11.10.**    Effect of Diffusion Parameter $D_e$ Variance on Breakthrough Curves in the Water System

such a correlation, we have defined a dimensionless diffusion coefficient $\overline{D}$, as follows:

$$\overline{D} = \frac{D_e V_b}{D_{AB} M r_p A_s} \tag{11.2}$$

where $V_b$ = pure sorbate molar volume, $D_{AB}$ = molecular diffusivity of sorbate in fluid phase, $A_s$ = total surface area per unit mass of adsorbent, $M$ = molecular weight of solute, $r_p$ = mean internal pore radius of sorbent.

This dimensionless diffusion coefficient will be correlated with a dimensionless equilibrium parameter, $\overline{E}$:

$$\overline{E} = \frac{\Delta C}{\Delta q \rho_s} \tag{11.3}$$

The relation between $\overline{D}$ and $\overline{E}$ has been derived as follows (Chapter III):

$$\overline{D} = D_0 \overline{E}^S \tag{11.4}$$

where $D_0$ and $S$ are the intercept and slope of a logarithmic plot of $\overline{D}$ and $\overline{E}$. Tables 11.12 and 11.13 contain the values of the parameters in Equation (11.4) for the sorbents and sorbates used in the correlation.

**Table 11.12.** Physical Parameters of Sorbents used in the Calculation of Generalized Correlation

| Adsorbent | $A_s$ (cm$^2$/g) | $r_p$ (cm) | $\rho_s$ (g/cm$^3$) |
|-----------|------------------|------------|---------------------|
| XAD-4 | $7.25 \times 10^6$ | $2 \times 10^{-7}$ | 0.694 |
| BAC | $1 \times 10^7$ | $3.25 \times 10^{-7}$ | 1.558 |
| MS-13X | $6.5 \times 10^6$ | $5 \times 10^{-8}$ | 1.428 |

Figure 11.11 shows a plot of $\overline{D}$ versus $\overline{E}$, for the liquid sorption data collected with the differential reactor column reported in Chapter III. The figure contains kinetic data on XAD-4 and XAD-2 resins. The correlation coefficient is 0.93.

Figure 11.12 presents the adsorption data for the organic compounds studied on XAD-4 resin (Chapter V), and shows that a systematic pattern exists for the gas phase sorption that is similar to the

**Table 11.13.**   Values [6] used in Calculation of Generalized
Correlation for Different Compounds

| Compound | MW (g/mole) | $V_b$ (cm³/mole) | $D_{AB}$ (cm²/sec) |
|---|---|---|---|
| **In Air:** | | | |
| $C_7H_8$ | 92.141 | 118.72 | $8.195 \times 10^{-2}$ |
| $C_8H_{10}$ | 106.168 | 143.24 | $7.485 \times 10^{-2}$ |
| $C_6H_6$ | 78.114 | 96.50 | $9.206 \times 10^{-2}$ |
| $CCl_4$ | 153.823 | 102.00 | $7.993 \times 10^{-2}$ |
| $C_2HCl_3$ | 131.389 | 95.21 | $8.641 \times 10^{-2}$ |
| $C_2Cl_4$ | 165.834 | 180.50 | $7.837 \times 10^{-2}$ |
| **In Water:** | | | |
| Phenol | 94.11 | 96.75 | $1.022 \times 10^{-5}$ |
| PCP | 128.56 | 112.82 | $0.932 \times 10^{-5}$ |

$V_b = 0.285V_c^{1.048}$. This equation was considered to be the least error method to estimate critical volume (except for benzene, $V_b$ value has already been well defined).

liquid systems. Table 11.14 lists the slope ($S$), and intercept ($D_0$) for all the systems reported in Figures 11.11 and 11.12. The slope for all of the adsorption systems are statistically indistinguishable and not statistically different from one another. Therefore, the tables also show the $D_0$ values when $S$ is set equal to 1.0. The intercept values ($D_0'$) shown in Table 11.14 are of considerable interest as they show much more variation than the slopes. For the gas phase data, the intercept varied for each sorbent-sorbate system. The variation due to different sorbents can be determined by comparing the average intercept value for individual sorbents. The values are 0.02 for the XAD-4 resin, 0.07 for the activated carbon and 0.90 for the 13X molecular sieve. The range of $D_0'$ values obtained for each sorbent (i.e., different sorbates) were 0.01 to 0.03 for the resin, 0.03 to 0.10 for the carbon and 0.2 to 1.6 for the sieve.

Molecular sieves are generally not used for adsorption of organic compounds. They are used in this study for comparison purposes and to test the dimensionless analyses method (i.e., small pore size).

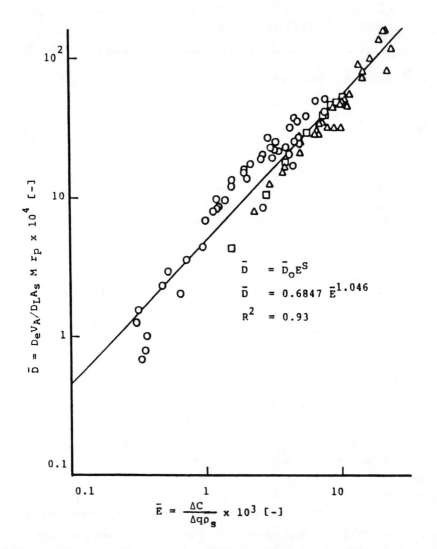

**Figure 11.11.** Dimensionless Effective Intraparticle Diffusion
Coefficient vs. Dimensionless Equilibrium Parameter

In Figure 11.13, data from the gas and liquid adsorption systems are
combined. The $\overline{D}$ values are divided by $D'_0$ so that all of the data can
be directly compared. A single line is obtained for the combined data
that has both a slope and intercept of one. The correlation coefficient
for the combined data is 0.98.

The relationship is based on data from 18 different gas-solid and
five different liquid-solid systems involving the adsorption of organic

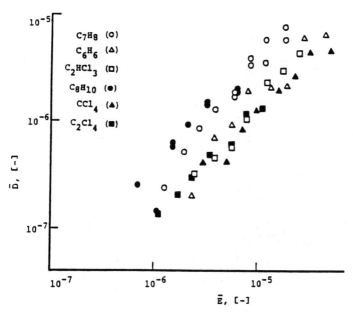

**Figure 11.12.**    Dimensionless Effective Diffusivity vs. Dimensionless Equilibrium Parameter for Adsorption of Organic Vapor on XAD-4

compounds with diverse physical properties onto resins, activated carbon, and molecular sieve. $D_e$ was converted to a dimensionless form by using information on the sorbate (molar volume, molecular diffusivity, and molecular weight) and the sorbent (surface area per mass of sorbate and pore radius). The water data have much higher dimensionless values than the air data, as shown by the fact that they are located on the upper part of the line in Figure 11.13.

Table 11.14.   Estimated Parameters for Generalized Correlation (Adsorption)

Correlation Equation: $\ln \bar{D} = \ln D_0 + S \ln \bar{E}$

| Adsorbent | Compound | $D_0 (\ln \sigma)^*$ | $S (\sigma)^*$ | $R^2$ |
|---|---|---|---|---|
| XAD-4 Resin | $C_7H_8$ | $9.47 \times 10^{-2}$ (0.85) | 1.12 ($8.82 \times 10^{-2}$) | 0.936 |
| | $C_8H_{10}$ | $7.17 \times 10^{-2}$ (1.71) | 1.07 ($1.61 \times 10^{-1}$) | 0.847 |
| | $C_6H_6$ | $3.56 \times 10^{-2}$ (1.20) | 1.09 ($1.30 \times 10^{-1}$) | 0.920 |
| | $CCl_4$ | $1.00 \times 10^{-2}$ (0.68) | 0.99 ($7.60 \times 10^{-2}$) | 0.966 |
| | $C_2HCl_3$ | $1.01 \times 10^{-1}$ (0.68) | 1.21 ($7.23 \times 10^{-2}$) | 0.982 |
| | $C_2Cl_4$ | $1.52 \times 10^{-2}$ (0.59) | 1.02 ($5.71 \times 10^{-2}$) | 0.985 |
| G-BAC Activated Carbon | $C_7H_8$ | $5.27 \times 10^{-2}$ (0.39) | 0.94 ($3.42 \times 10^{-2}$) | 0.991 |
| | $C_8H_{10}$ | $2.06 \times 10^{-1}$ (0.42) | 1.04 ($3.45 \times 10^{-2}$) | 0.995 |
| | $C_6H_6$ | $9.15 \times 10^{-3}$ (1.08) | 0.85 ($9.41 \times 10^{-2}$) | 0.931 |
| | $CCl_4$ | $2.71 \times 10^{-2}$ (0.45) | 0.93 ($3.93 \times 10^{-2}$) | 0.989 |
| | $C_2HCl_3$ | $2.27 \times 10^{-2}$ (0.53) | 0.95 ($4.56 \times 10^{-2}$) | 0.986 |
| | $C_2Cl_4$ | $4.59 \times 10^{-2}$ (0.01) | 1.00 ($1.10 \times 10^{-2}$) | 0.999 |
| MS-13X Molecular Sieve | $C_7H_8$ | 2.11 (0.74) | 1.09 ($7.37 \times 10^{-1}$) | 0.976 |
| | $C_8H_{10}$ | 3.28 (0.48) | 1.09 ($4.40 \times 10^{-2}$) | 0.992 |
| | $C_6H_6$ | 0.50 (0.49) | 0.99 ($5.77 \times 10^{-2}$) | 0.990 |
| | $CCl_4$ | 0.30 (0.37) | 1.07 ($3.73 \times 10^{-2}$) | 0.995 |
| | $C_2HCl_3$ | 0.75 (0.47) | 1.16 ($5.55 \times 10^{-2}$) | 0.991 |
| Water-Solid System | | | | |
| XAD-4 PCP | | 1.00 (0.18) | 1.05 ($3.12 \times 10^{-2}$) | 0.932 |
| XAD-2 Phenol | | | | |

Correlation Equation: $\bar{D} = D'_0 \bar{E}$ (when $S = 1.0$)

| Adsorbent | Compound | $D'_0$ | $(\sigma)$ | $R^2$ |
|---|---|---|---|---|
| XAD-4 Resin | $C_7H_8$ | $2.96 \times 10^{-2}$ | $(2.53 \times 10^{-3})$ | 0.919 |
| | $C_8H_{10}$ | $3.28 \times 10^{-2}$ | $(1.87 \times 10^{-3})$ | 0.972 |
| | $C_6H_6$ | $1.60 \times 10^{-2}$ | $(1.34 \times 10^{-3})$ | 0.953 |
| | $CCl_4$ | $1.06 \times 10^{-2}$ | $(7.34 \times 10^{-4})$ | 0.967 |
| | $C_2HCl_3$ | $1.68 \times 10^{-2}$ | $(6.56 \times 10^{-4})$ | 0.991 |
| | $C_2Cl_4$ | $1.25 \times 10^{-2}$ | $(5.77 \times 10^{-4})$ | 0.87 |
| G-BAC Activated Carbon | $C_7H_8$ | $8.30 \times 10^{-2}$ | $(7.28 \times 10^{-3})$ | 0.942 |
| | $C_8H_{10}$ | $1.33 \times 10^{-1}$ | $(4.17 \times 10^{-3})$ | 0.994 |
| | $C_6H_6$ | $3.06 \times 10^{-2}$ | $(7.08 \times 10^{-3})$ | 0.727 |
| | $CCl_4$ | $5.30 \times 10^{-2}$ | $(1.86 \times 10^{-3})$ | 0.991 |
| | $C_2HCl_3$ | $3.17 \times 10^{-2}$ | $(2.62 \times 10^{-3})$ | 0.955 |
| | $C_2Cl_4$ | $4.33 \times 10^{-2}$ | $(6.51 \times 10^{-4})$ | 0.999 |
| MS-13X Molecular Sieve | $C_7H_8$ | 1.04 | $(2.51 \times 10^{-2})$ | 0.996 |
| | $C_8H_{10}$ | 1.57 | $(7.97 \times 10^{-2})$ | 0.985 |
| | $C_6H_6$ | 0.53 | $(1.61 \times 10^{-2})$ | 0.996 |
| | $CCl_4$ | 0.18 | $(1.82 \times 10^{-3})$ | 0.999 |
| | $C_2HCl_3$ | 0.26 | $(1.16 \times 10^{-2})$ | 0.990 |
| Water-Solid System | | | | |
| XAD-4 PCP | | 0.55 | $(1.41 \times 10^{-2})$ | 0.949 |
| XAD-2 Phenol | | | | |

* $\sigma$ = Standard deviation.

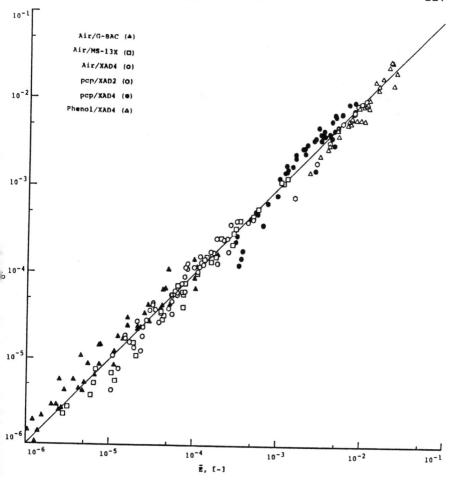

**Figure 11.13.**    General Correlation for the Effective Diffusion
Coefficient in Gas and Liquid Systems ($n = 216, r^2 = 0.98$)

*REFERENCES*

1. Wilhoit, T., and R. Zwolinski, *Handbook of Vapor Pressures and Heats of Vaporization of Hydrocarbons and Related Compounds*, American Petroleum Institute, Evans Press, Fortworth, TX.

2. Hand, D.W., J.C. Crittenden, and W.E. Thacker, *ASCE. EE.*, **110**, 440 (1984).

3. Villadsen, J.V., and W.E. Stewart, *Chem. Eng. Sci.*, **22**, 1483 (1967).

4. Sarlis, J.N., *Sorption Behavior of Hazardous Organic Solvents by Direct Differential Reactor Measurement*, M.S. Thesis, Illinois Institute of Technology, Chicago, IL (1985).

5. Yeh, M.C., *Modeling of Organic Vapors Adsorption in Single Component Systems*, Ph.D. Thesis, Illinois Institute of Technology, Chicago, IL. (1987).
6. Reid, R.C., J.M. Prausnitz, and T.K. Sherwood, *The Properties of Gases and Liquids*, 3rd ed., McGraw-Hill Book Company, New York (1977).

# CHAPTER XII

---

## THERMODYNAMIC ANALYSIS
## OF ADSORPTION SYSTEMS

---

Contributing Author: M.C. Yeh

## INTRODUCTION

The rate of adsorption in porous adsorbents is generally controlled by transport within the pore network. The intraparticle transport is considered a diffusion process taking place by several different mechanisms (ordinary diffusion or molecular diffusion, Knudsen diffusion and surface diffusion). Their relative contribution depends on the pore size, the sorbate concentration, and other conditions.

Surface diffusion plays an important role in the transport of gases through beds of porous and non-porous solids. Carman [1], Rutz and Kammermeyer [2], and Smith and Metzner [3] have shown surface migration rates generally to increase with adsorbate concentration. This invalidates the simple concept of a diffusion process having a constant diffusion coefficient. The increase of surface diffusivity with surface concentration can be interpreted in terms of several mechanisms [4]: (1) the increase of unit hopping distance for the hopping mode, (2) the change in the ratio of the chemical potential gradient to the gradient of the amount adsorbed with the increase of surface concentration, (3) the decrease in the heat of adsorption with increased concentration.

In this chapter, surface diffusivity in a spherical particle will be correlated with surface coverage and heat of adsorption. Rate data from a differential bed reactor were analyzed using a combined diffusion model which considered both pore and surface diffusion within the adsorbent particle. A numerical technique was employed for the solution of the diffusion equation. Experiments were conducted at three different temperatures, ($25^0$C, $50^0$C, and $75^0$C) to assess the effect of temperature on the surface diffusion coefficient and the adsorption equilibrium isotherms.

## METHOD AND RESULTS

### Models and Method

Combined Diffusion Model. The internal diffusion can be expressed by two possible simultaneous mechanisms of diffusion: pore diffusion, including molecular or Knudsen diffusion, and surface diffusion [5,6]. If one considers this combined parallel resistance within the adsorbent particle, a material balance for spherical particle can be expressed by Equation (2.46). The boundary and initial conditions are Equations (2.47)–(2.49).

Pore Diffusivity. The mass flux through the gas phase within the pore, expressed by $D_p$ as Equation (2.38), is the result of both the Knudsen and molecular diffusion mechanisms [7]. The effective molecular diffusivity, $D_{m,eff}$, can be calculated from the molecular diffusivity, $D_m$, given by the Chapman-Enskog equation [8] corrected for the tortuosity factor, $\tau$, and the porosity of the particle, as shown by Equations (2.33)–(2.35). Kinetic theory provides Equation (2.37) for effective Knudsen diffusivity, $D_{k,eff}$, in gas-solid systems. The tortuosity factor used in those equations is a characteristics of the geometry of the porous adsorbent. As such, it should be independent of the diffusing gas and operating conditions.

Surface Diffusivity. The surface diffusion coefficient is dependent on the adsorbed concentration and temperature. This concentration dependency can be interpreted using the hopping model. On a homogeneous surface, the surface flow phenomenon is regarded as a random walk process on the surface. Since the activity of each active site is the same, the probability and distance of hopping to another vacant site is a function of surface coverage. Therefore, the surface diffusivity increases when surface coverage is higher.

Higashi et al. [9] and Okazaki et al. [10] have proposed a hopping model for homogeneous surfaces. The model assumed that every molecule occupies a single site and jumps over a potential barrier one by one. An equation describing this relation was derived as follows:

$$D_s = \frac{1}{1-x} D_0 \exp(-E/RT) \tag{12.1}$$

where $x$ = surface coverage; $D_0$ = surface diffusivity for zero surface coverage; $E$ = activation energy and $R$ = gas constant.

For adsorption on an energetically heterogeneous surface, the surface transport is affected by the distribution of binding energy to the surface. The heat of adsorption for a gas-solid system evidently decreases with solid phase concentration because of the progressive filling of sites of decreasing strength. The variation in surface diffusivity with surface coverage is due to progressive filling of sites of decreasing energy.

Gilliand *et al.* [11] and Sladek [12] described the surface transport in terms of hopping of adsorbed molecules between adjacent sites of different adsorption strengths:

$$D_s = D_0 \exp(-a\Delta H/RT) \qquad (12.2)$$

and

$$D_s = D_0 \exp(-E/RT) \qquad (12.3)$$

where $a$ = constant and $\Delta H$ = heat of adsorption.

*Experimental Procedure*

Adsorbents. The two adsorbents used in this study were beaded activated carbon (BAC) manufactured by Union Carbide and a resin, XAD-4, made synthetically by Rohm and Haas Co.. The physical properties of the adsorbents are listed in Table 1.1. In an attempt to achieve a more uniform size, BAC was sieved down to 20–30 mesh size. The average particle size was 0.0674 cm. For resin XAD-4, a size fraction which passed a 30 mesh sieve and was retained on a 40-mesh sieve was collected. The mean particle diameter was 0.046 cm.

Adsorbates. The adsorbates, toluene ($C_7H_8$), $p$-xylene ($p$-$C_8H_{10}$), benzene ($C_6H_6$), carbon tetrachloride ($CCl_4$), tetrachloroethylene ($CCl_2$= $CCl_2$) and trichloroethylene ($C_2HCl_3$), were carried by nitrogen gas ($N_2$) during the experiments. Table 4.1 shows the physical properties of these adsorbates.

Procedure. The experimental system was essentially the same as the one used previously by Noll *et al.* [13–15]. The detailed procedure was described in Chapter IV. However, the rate data were obtained at different temperatures, i.e., 25°C, 50°C, and 75°C. The experimental concentration-time data were compared to the kinetic curve predicted by the differential bed reactor model and the best statistical description determined the kinetic parameters.

*Method of Approach*

Model of Differential Reactor. The model assumes: (1) the adsorbent particle is at isothermal state, i.e., there is no temperature gradient within the particle, (2) fluid phase concentration remains constant [16], (3) external mass transfer resistance can be neglected [17], (4) surface diffusivity is constant for each constant input vapor concentration, (5) local equilibrium exists and the Freundlich isotherm can be applied. Then Equation (6.7), coupled with boundary and initial conditions, Equations (6.8)–(6.10), was used to describe the rate of mass transfer within the adsorbent particle and to determine the intraparticle diffusion coefficient. The solution to Equation (6.7) utilized the orthogonal collocation method [18]. After numerical testing, $Nc = 7$ was shown to be sufficient to give accurate results; therefore, collocation number seven was applied in the analyses.

## Results

Equilibrium Isotherm. The Freundlich isotherm was able to express the equilibrium conditions of each system for each temperature. The parameters of each particular isotherm listed in Table 12.1 were obtained by a statistical analysis of the gravimetric equilibrium data.

Heat of Adsorption. The isosteric heat of adsorption $\Delta H$ was determined from the equilibrium data using the isosteric equation:

$$\Delta H = R \frac{\partial \ln P}{\partial(1/T)}\bigg|_q \qquad (12.4)$$

where $P$ is the vapor pressure. A plot of $\ln P$ versus $1/T$ can be analyzed graphically for the slope at constant loading amount to obtain $\Delta H$. Typical isosteres are shown in Figures 12.1 and 12.2. The isosteric heat of adsorption, as a function of $q$, is illustrated in Figures 12.3 and 12.4 for some systems.

$D_s$, $D_p$, $\tau$ Determination. The solution of the model yields a curve of normalized uptake, $\overline{q}_{av}$, versus dimensionless time, $\overline{T}$, for a given value of $D_s/D_p$. That predicted curve is compared with the experimental curve of $q$ versus $t$, which will give a best fitting value of $D_s/D_p$. The value so derived is effective for that specific concentration range.

Equation (6.11) suggests that if $\overline{T}$ is plotted versus $t$, the line should pass exactly through the given point (0,0) and with a slope of $D_p/R_p^2$. Prahl [19] pointed out that the slope of the least-square line through a given point (0,0) can be calculated using Equation (12.5).

$$b_s = \frac{\sum t_i \overline{T}_i}{\sum t_i^2} \qquad (12.5)$$

where $b_s$ = slope of the least-square line; $t_i$ = various experimental time and $\overline{T}_i$ = corresponding dimensionless time. A detailed procedure for the determination of $D_s$ and $D_p$ at $25^0\mathrm{C}$ was explained in Chapter VI.

In Figures 12.5–12.16, the $D_s$ value for some systems are plotted versus the adsorbed concentration at different temperatures. A strong

**Table 12.1.** Freundlich Parameters* for Different Systems at Different Temperatures

| System | $T(^0C)$ | $K$ | $n$ | $r^2$ |
|---|---|---|---|---|
| $C_6H_6$/XAD-4 | 25 | 2.06 | 2.13 | 0.98 |
| | 50 | 0.188 | 1.536 | 0.94 |
| | 75 | 0.150 | 1.635 | 0.99 |
| $C_6H_6$/BAC | 25 | 62.01 | 5.26 | 0.93 |
| | 50 | 29.01 | 3.794 | 0.98 |
| | 75 | 17.53 | 3.547 | 1.00 |
| $C_7H_8$/XAD-4 | 25 | 3.37 | 1.936 | 0.99 |
| | 50 | 1.22 | 1.922 | 0.99 |
| | 75 | 0.454 | 1.803 | 1.00 |
| $C_7H_8$/BAC | 25 | 178.41 | 11.74 | 0.96 |
| | 50 | 100.22 | 7.23 | 0.98 |
| | 75 | 51.66 | 4.95 | 0.98 |
| $p$-$C_8H_{10}$/XAD-4 | 25 | 7.16 | 1.96 | 0.99 |
| | 50 | 3.182 | 2.036 | 1.00 |
| | 75 | 1.451 | 1.957 | 1.00 |
| $p$-$C_8H_{10}$/BAC | 25 | 258.72 | 20.2 | 0.72 |
| | 50 | 175.67 | 12.374 | 0.97 |
| | 75 | 121.86 | 8.425 | 0.99 |
| $CCl_4$/XAD-4 | 25 | 2.51 | 2.0 | 0.99 |
| | 50 | 0.706 | 1.797 | 1.00 |
| | 75 | 0.491 | 1.869 | 1.00 |
| $CCl_4$/BAC | 25 | 141.57 | 6.25 | 0.95 |
| | 50 | 48.21 | 3.682 | 0.98 |
| | 75 | 18.8 | 2.906 | 1.00 |
| $C_2HCl_3$/XAD-4 | 25 | 2.88 | 2.0 | 0.99 |
| | 50 | 0.918 | 1.857 | 1.00 |
| | 75 | 0.26 | 1.643 | 1.00 |
| $C_2HCl_3$/BAC | 25 | 158.42 | 6.67 | 0.94 |
| | 50 | 70.079 | 4.463 | 0.98 |
| | 75 | 27.483 | 3.265 | 0.99 |

*(continued)*

Table 12.1.   Freundlich Parameters* for Different Systems at Different Temperatures *(continued)*

| System | $T(^{0}C)$ | $K$ | $n$ | $r^2$ |
|---|---|---|---|---|
| $C_2Cl_4$/XAD-4 | 25 | 7.63 | 2.0 | 0.99 |
| | 50 | 2.01 | 1.82 | 1.00 |
| | 75 | 0.182 | 1.35 | 0.99 |
| $C_2Cl_4$/BAC | 25 | 353.25 | 11.11 | 0.94 |
| | 50 | 217.31 | 7.66 | 0.99 |
| | 75 | 99.75 | 5.13 | 0.99 |

* $C_e$ is in ppm and $q_e$ in mg/g.

increase of $D_s$ with an adsorbed amount and temperature was exhibited by all the adsorbent/adsorbate systems studied.

### $D_s$ Variation and Surface Character

Activated Carbon Systems. Figure 12.3 shows that the isosteric heat of adsorption decreases significantly as the adsorbed amount increases for the carbon adsorption systems. This indicates that activated carbon has a heterogeneous surface. The concentration dependency of the diffusion coefficient can be attributed to the changes in the strength of adsorption. On the basis of a model accounting for the hopping of adsorbed molecules between sites of different energy, a strong variation of $D_s$ with the isosteric heat of adsorption is expected. A correlation of surface diffusivity with the heat of adsorption in terms of Equation (12.2) (for one BAC system) is shown in Figure 12.17. By this means, an $a$ value and a $D_0$ value can be estimated from the slope and the intercept of the line on the log $D_s$-axis, respectively. Similar correlations were obtained for the rest of the BAC systems, as shown in Figures 12.18–12.22. The constants   *Do, a,*   and $r^2$ from each correlation are summarized in Table 12.2. The ratio of activation energy to the heat of adsorption, $a = E/\Delta H$, was found to be nearly constant for each system. Tabulated values show that the activation energy was between 0.2 and 0.37 of the heat of adsorption for these systems.

XAD-4 Systems. Figure 12.4 shows the isosteric heat of adsorption for the six selected systems. For carbon-chloride compounds, the in-

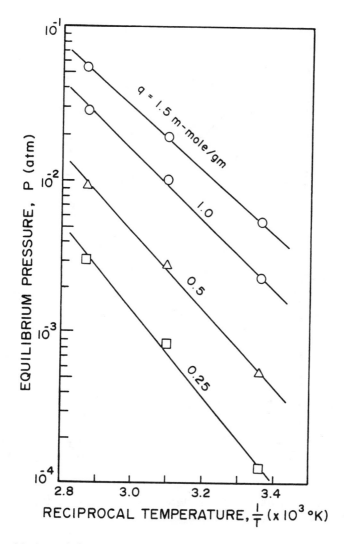

**Figure 12.1.**    Adsorption Isosteres for Determining Isosteric
   Heat of Adsorption ($C_6H_6$/XAD-4)

creasing energetic heterogeneity is in the sequence of $CCl_4$, $C_2HCl_3$
and $C_2Cl_4$. In the case of carbon tetrachloride (non-polar), the near
constant value for the heat of adsorption suggests that the surface is
almost energetically uniform. For the more strongly polar molecules,
the decrease in the heat of adsorption with surface coverage becomes
more pronounced. The molecules of trichloroethylene have a stronger
polar character [20] and show this tendency. Although another halo-
genated hydrocarbon, tetrachloroethylene, is a non-polar compound,

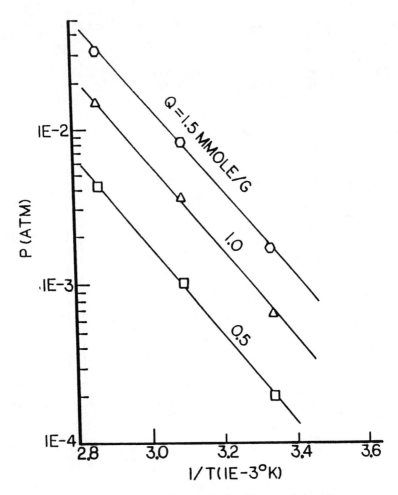

**Figure 12.2.**    Adsorption Isosteres for Determining Isosteric
Heat of Adsorption ($C_7H_8$/XAD-4)

the effect of the double bond may give rise to an additional electrostatic contribution to the heat of adsorption.

The heat of adsorption of toluene and p-xylene are almost constant
over a wide range of concentration.  On the other hand, the heat of
adsorption of benzene is strongly dependent on the surface coverage.
Both toluene and p-xylene are formed by substitution of an alkyl group
on an aromatic ring, whereas benzene is not substituted.  The substitution of a methyl group on the aromatic ring increases the molecular
size and decreases the electrostatic effect of $\pi$-bond on the aromatic
nature of the adsorbent surface.

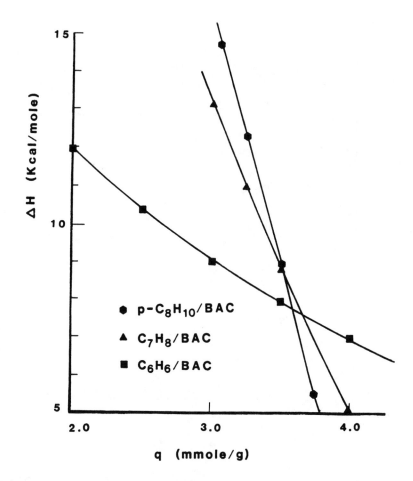

**Figure 12.3.** Variation of Heat of Adsorption with Solid Phase Concentration for Carbon BAC

The XAD-4 surface appears almost energetically uniform to the large molecules ($C_7H_8$ and $p$-$C_8H_{10}$) and to the non-polar molecule ($CCl_4$), and shows slight heterogeneity for the polar molecule ($C_2HCl_3$). This behavior (the extent of energetical heterogeneity of an adsorbent surface depends on the polarity and the size of the sorbate molecule) has been found in other systems [21,22]. However, the heterogeneous character shown by non-polar molecules, $C_6H_6$ and $C_2Cl_4$, might result from more complicated sorbate-sorbate and sorbate-sorbent interactions.

The correlations of surface diffusion in terms of Equation (12.2) (heterogeneous surface) for $C_2Cl_4$, $C_2HCl_3$ and $C_6H_6$ are shown in

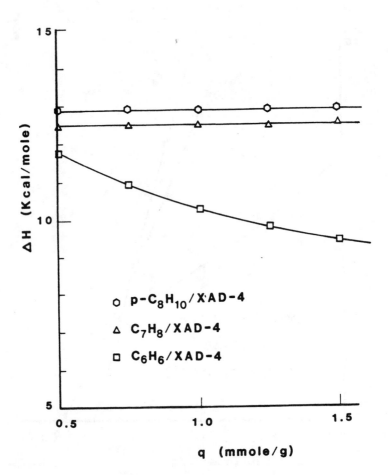

**Figure 12.4.**    Variation of Heat of Adsorption with Solid Phase
Concentration for Resin XAD-4

Figures 12.23–12.25. Table 12.2 shows the constants $D_0$, $a$, and $r^2$ for
each system. The ratio of activation energy to the heat of adsorption
was between 0.48–0.52.

On a homogeneous surface, Equations (12.2) and (12.3) cannot explain the dependency of $D_s$ on the amount adsorbed, $q$, because the
heat of adsorption $\Delta H$ in Equation (12.2) does not change with $q$, as
shown in Figure 12.4. Therefore, corrected diffusivities, $D_c$, calculated
from Equation (12.7) were essentially independent of concentration
for the adsorption of homogeneous systems.

$$Dc = D_s(1 - x) \tag{12.6}$$

$$Dc = D_0 \exp(-a\Delta H / RT) \tag{12.7}$$

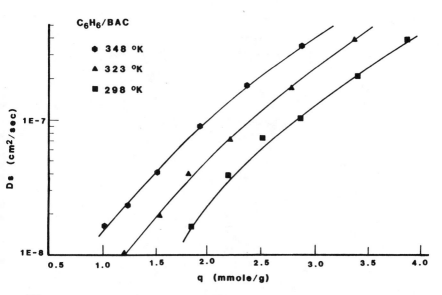

Figure 12.5.   $D_s$ vs. $Q$ at Different Temperature for $C_6H_6$/BAC

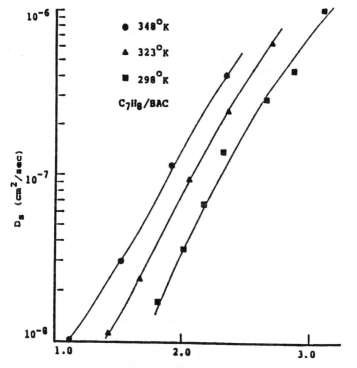

Figure 12.6.   $D_s$ vs. $Q$ at Different Temperature for $C_7H_8$/BAC

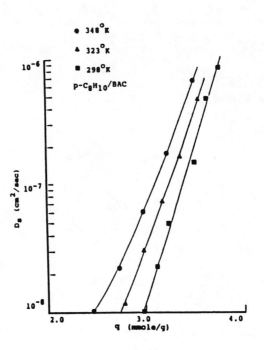

**Figure 12.7.** $D_s$ vs. $Q$ at Different Temperature for $p$-$C_8H_{10}$/BAC

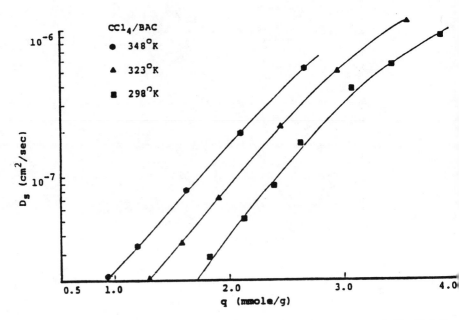

**Figure 12.8.** $D_s$ vs. $Q$ at Different Temperature for $CCl_4$/BAC

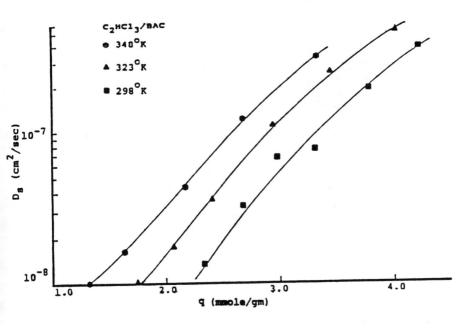

**Figure 12.9.** $D_s$ vs. $Q$ at Different Temperature for $C_2HCl_3/BAC$

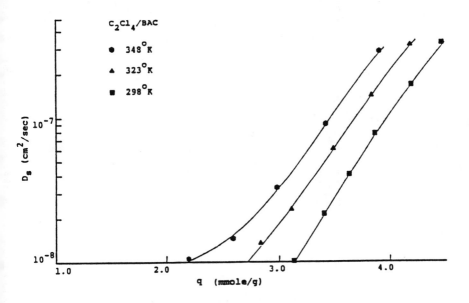

**Figure 12.10.** $D_s$ vs. $Q$ at Different Temperature for $C_2Cl_4/BAC$

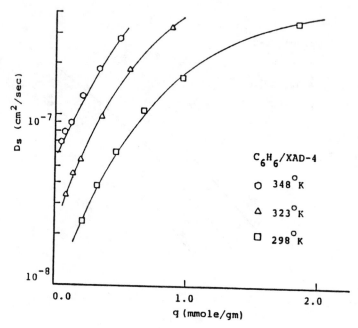

**Figure 12.11.**    $D_s$ vs. $Q$ at Different Temperature for $C_6H_6$/XAD-4

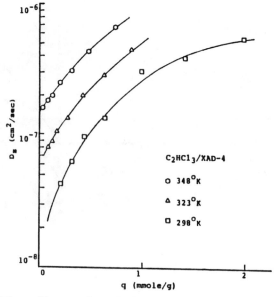

**Figure 12.12.**    $D_s$ vs. $Q$ at Different Temperature for $C_2HCl_3$/XAD-4

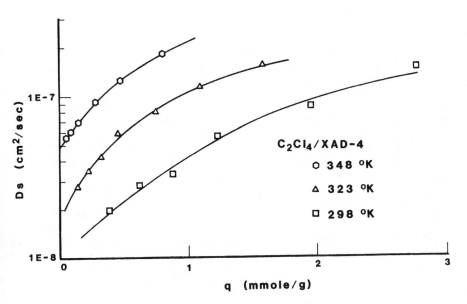

**Figure 12.13.**    $D_s$ vs. $Q$ at Different Temperature for
$C_2Cl_4$/XAD-4

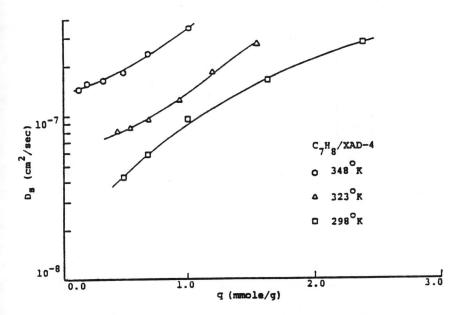

**Figure 12.14.**    $Q_s$ vs. $Q$ at Different Temperature for
$C_7H_8$/XAD-4

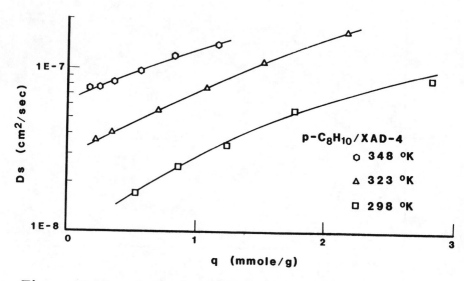

**Figure 12.15.**    $Q_s$ vs. $Q$ at Different Temperature for $p$-$C_8H_{10}$/XAD-4

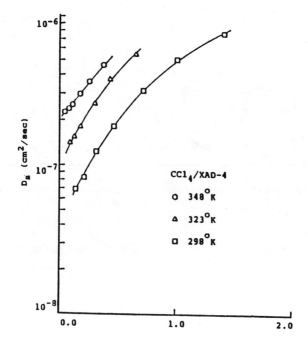

**Figure 12.16.**    $Q_s$ vs. $Q$ at Different Temperature for $CCl_4$/XAD-4

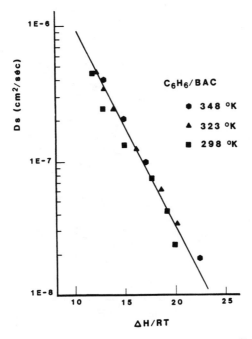

**Figure 12.17.**    Diffusivity Correlation for $C_6H_6$/BAC

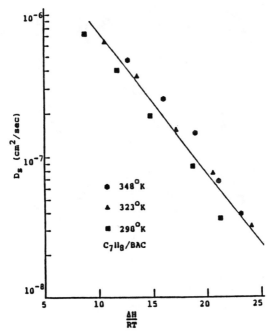

**Figure 12.18.**    Diffusivity Correlation for $C_7H_8$/BAC

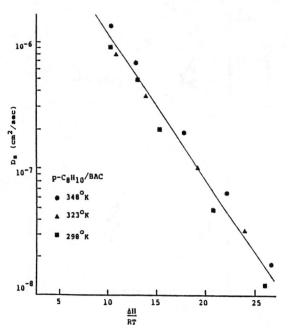

**Figure 12.19.**    Diffusivity Correlation for $p$-$C_8H_{10}$/BAC

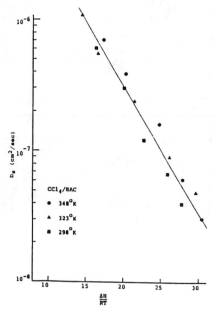

**Figure 12.20.**    Diffusivity Correlation for $CCl_4$/BAC

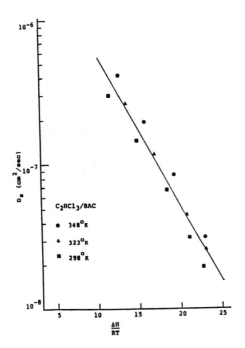

**Figure 12.21.**     Diffusivity Correlation for $C_2HCl_3/BAC$

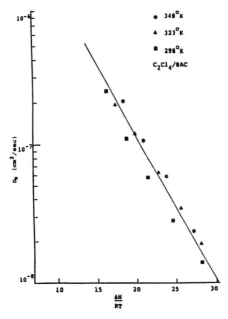

**Figure 12.22.**     Diffusivity Correlation for $C_2Cl_4/BAC$

**Table 12.2.** $D_s$ Correlation Parameters for Different Systems

| System (Character)* | $q$ (mmole/g) | $-\Delta H$ (Kcal/mole) | $-a$ (−) | $D_0 \times 10^5$ (cm$^2$/sec) | $r^2$ (−) |
|---|---|---|---|---|---|
| $CCl_4$/BAC (Heterogeneity) | 1.50-3.50 | 6.8-15.4 | 0.219 | 0.792 | 0.94 |
| $C_2HCl_3$/BAC (Heterogeneity) | 2.25-4.00 | 6.8-14.8 | 0.253 | 0.783 | 0.99 |
| $C_2Cl_4$/BAC (Heterogeneity) | 3.00-4.50 | 7.3-18.0 | 0.234 | 1.18 | 0.93 |
| $C_6H_6$/BAC (Heterogeneity) | 1.50-4.00 | 7.0-13.9 | 0.372 | 4.54 | 0.97 |
| $C_7H_8$/BAC (Heterogeneity) | 2.75-4.00 | 5.0-15.0 | 0.244 | 1.03 | 0.95 |
| $p$-$C_8H_{10}$/BAC (Heterogeneity) | 3.00-3.75 | 5.5-15.4 | 0.261 | 1.50 | 0.97 |
| $CCl_4$/XAD-4 (Homogeneity) | 0.15-1.50 | 10.1-11.5 | 0.48 | 30.7 | 0.99 |
| $C_2HCl_3$/XAD-4 (Heterogeneity) | 0.10-2.00 | 9.8-13.9 | 0.52 | 311 | 0.99 |
| $C_2Cl_4$/XAD-4 (Heterogeneity) | 0.10-2.00 | 10.5-16.2 | 0.52 | 249 | 0.99 |
| $C_6H_6$/XAD-4 (Heterogeneity) | 0.10-1.50 | 9.4-12.7 | 0.48 | 89.4 | 0.98 |
| $C_7H_8$/XAD-4 (Homogeneity) | 0.12-3.47 | 12.5 | 0.45 | 49.9 | 0.99 |
| $p$-$C_8H_{10}$/XAD-4 (Homogeneity) | 0.17-5.74 | 12.9 | 0.51 | 103 | 0.99 |

* Surface Character

in which the values of $x$ were obtained from the original equilibrium data.

Figures 12.26–12.28 show the corrected surface diffusivity versus solid phase concentration at different temperatures for the systems of $p$-$C_8H_{10}$/XAD-4, $C_7H_8$/XAD-4 and $CCl_4$/XAD-4. The tempera-

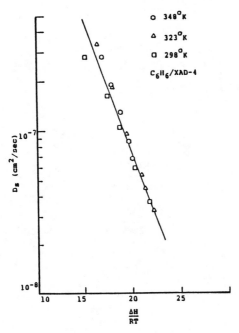

**Figure 12.23.** Diffusivity Correlation for $C_6H_6$/XAD-4

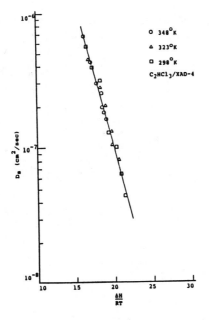

**Figure 12.24.** Diffusivity Correlation for $C_2HCl_3$/XAD-4

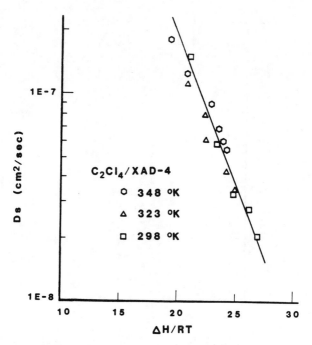

**Figure 12.25.** Diffusivity Correlation for $C_2Cl_4$/XAD-4

ture dependency of the corrected diffusivity correlated from Equation (12.7) is shown in Figure 12.29. The values of $D_0$ and $a$ are shown in Table 12.2. The ratio of activation energy to the heat of adsorption for these homogeneous systems varied between 0.45 and 0.51.

In the present study, the value of parameter $a$ does not differ greatly for various adsorbates: 0.21–0.37 for activated carbon and 0.45–0.52 for XAD-4. Gilliland et al. [11] proposed that $a$-values range from 0.43 to 0.8 for physical adsorbed materials and from 0.15 to 0.25 for chemisorbed materials. The results in the present work are in good agreement with those values. The parameter $D_0$ in Equations (12.2) or (12.7) is of the order of $10^{-4}$–$10^{-3}$ cm$^2$/sec for XAD-4 and $10^{-6}$–$10^{-5}$ cm$^2$/sec for activated carbon. Compared with the $D_0$-values presented by Okazaki and his co-workers [10]: $10^{-3}$–$10^{-2}$ cm$^2$/sec for porous Vycor glass, Linda-silica, Carbolac and Carbon Regal, and $10^{-1}$ cm$^2$/sec for Graphon, the results for different solids are much different.

*CONCLUSIONS*

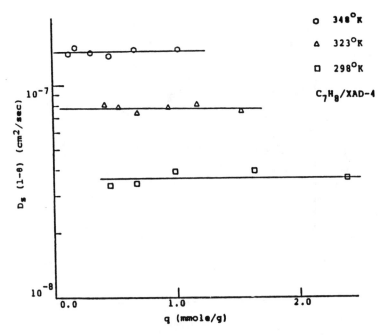

**Figure 12.26.**    $D_c$ vs. $Q$ at Different Temperature for
$C_7H_8$/XAD-4

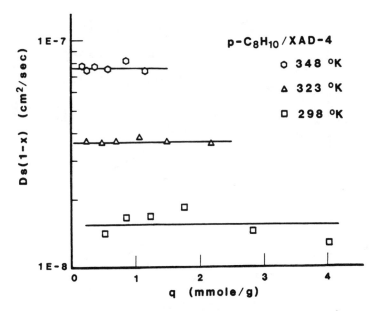

**Figure 12.27.**    $D_c$ vs. $Q$ at Different Temperature for
$p$-$C_8H_{10}$/XAD-4

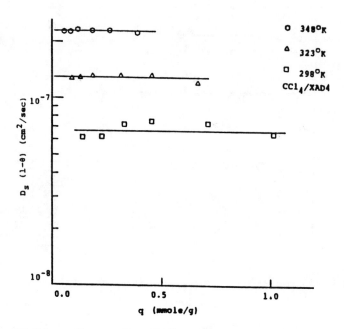

**Figure 12.28.**    $D_s$ vs. $Q$ at Different Temperature for
CCl$_4$/XAD-4

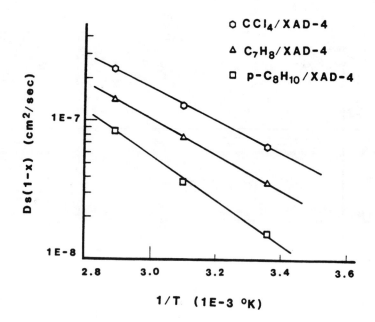

**Figure 12.29.** Temperature Dependency of Corrected Diffusivities

In summary, findings of this research are:

1. The kinetic data of the differential bed reactor can be well described by the combined diffusion model. This method provides an approach procedure to evaluate the diffusivities at different concentrations and temperatures for different adsorption systems.

2. The surface diffusion coefficient is highly dependent on the surface character, concentration, and temperature.

3. The activated carbon has an energetically heterogeneous surface.

4. The extent of the energetical heterogeneity of XAD-4 is dependent on the chemical and physical properties of the adsorbate.

5. The ratio of activation energy to the isosteric heat of adsorption is almost constant for the adsorbent with various adsorbates.

6. The parameter $D_0$ varied for different adsorbate/adsorbent systems. This makes the correlation between surface diffusivity and heat of adsorption system specific. More investigation of $a$ and $D_0$ may eventually establish a general procedure for the prediction of the surface diffusion coefficient.

## REFERENCES

1. Carman, P.C. *Flow of Gases Through Porous Media*. Academic Press, New York, NY (1956).
2. Rutz, P.O., and K. Kammermeyer, *U.S. Atomic Energy Commission Report AECU-4328, Flow Through Microporous Media – Vapor Transfer Through Barriers*. n.p.
3. Smith, R.K., and A.B. Metzner, *J. Phys. Chem.*, **68**, 2741 (1964).
4. Sudo, Y., and M. Suzuki, *Chem. Eng. Sci.*, **33**, 1287 (1977).
5. Costa, E., G. Calleja, and F. Domingo, *AIChE. J.*, **31**, 982 (1985).
6. Sheindorf, C., M. Rebhun, and M. Sheintuch, *Chem. Eng. Sci.*, **38**, 335 (1983).
7. Pollard, W.G., and R.D. Present, *Chem. Rev.*, **73**, 762 (1948).
8. Hirshfelder, J.O., C.F. Curtiss, and R.B. Bird, *Molecular Theory of Gases and Liquids*. John Wiley & Sons, New York, NY (1954).
9. Higashi, K., H. Ito, and J. Oishi, *J. Atomic Energy Soc. Japan*, **5**, 846 (1963).
10. Okazaki, M., H. Tamon, and R. Toei, *AIChE. J.*, **27**, 262 (1981).
11. Gilliland, E.R., R.F. Baddour, G.P. Perkinson, and K.J. Sladek, *Ind. Eng. Chem. Fundam.*, **13**, 95 (1974).
12. Sladek, K.J., E.R. Gilliland, and R.F. Baddour, *Ind. Eng. Chem. Fundam.*, **13**, 100 (1974).
13. Noll, K.E., C.N. Haas, A.A. Aguwa, M. Satoh, A. Belalia, and P.S. Bartolomew, *Direct Differential Reactor Studies on Adsorption from Industrial Strength Liquid Gaseous Solutions*. Presented at the Engineering Foundation Conference, May 1983, at Schloss Elmau, Bavaria, West Germany.
14. Noll, K.E., and J.N. Sarlis, *J. APCA.*, **38**, 1512 (1988).
15. Noll, K.E., A.A. Aguwa, Y.P. Fang, and P.T. Boulanger, *ASCE. EE.*, **111**, 487 (1985).
16. Fang, Y.P. *Adsorption and Desorption Study of Toluene Vapor on XAD-4 Using Single Particle Layer Method*. M.S. Thesis, Illinois Institute of Technology, Chicago, IL (1984).
17. Carlson, N.W., and J.S. Dranoff, *Ind. Chem. Process Des. Dev.*, **24**, 1300 (1985).
18. Finlayson, B.A. *The Method of Weighted Residuals and Variational Principles*. Academic Press, New York, NY (1972).
19. Prahl, W.H. *Chem. Eng.*, **5**, 65 (1983).
20. Carey, F.A., and R.J. Sundbery, *Advance Organic Chemistry*. Plenum Press, New York, NY (1977).
21. Ruthven, D.M. *Principles of Adsorption and Adsorption Processes*. John Wiley & Sons, New York, NY (1984).

2. Ruthven, D.M. *AIChE. Symp.*, **80**, 21 (1984).

# CHAPTER XIII

## APPLICATION OF POTENTIAL THEORY
## FOR GAS ADSORPTION SYSTEMS

Contributing Authors: T. Shen and D.H. Wang

## *INTRODUCTION*

The adsorption isotherm models presented in Chapter II (Freundlich, Langmuir, etc), are generally able to describe the relationship between the adsorbate concentrations in the fluid and adsorbed phases. Those descriptions are based on experimental data that are used to calculate parameters of the model by statistical curve fitting. The parameters of those models, however, are system and temperature specific. This means that a separate experimental study has to be conducted, prior to any particular application of the adsorption process, to determine the isothermal parameters for the specific system and at the given temperature. If more than one adsorbent is to be considered, the entire experimental procedure has to be repeated separately for each adsorbent.

A much more simple and convenient approach to adsorption equilibria is offered by the adsorption potential theory. This theory predicts, in principle, the adsorption capacity of a given adsorbent for any adsorbate at any concentration and at any temperature. These predictions are based on a single isotherm of a reference adsorbate constructed experimentally, once, for the particular adsorbent.

The description above represents the ideal case. In practice, and at the current state of experience with the method, some complications occur requiring the use of more than one reference adsorbate, and the application of special normalizing techniques to extrapolate a reference isotherm to other systems. Those complications can be handled and they do not limit the applicability and value of this method, which when properly applied, provides predictions sufficiently accurate for design purposes. The theory and application of this method are given in this chapter, along with an evaluation of alternative approaches to deal with deviations from the ideal case.

## *METHOD AND RESULTS*

*Theory*

The adsorption potential theory primarily describes physical adsorption onto a microporous solid. It postulates that the phenomenon of adsorption takes place in a space of fixed volume, close to the surface of the adsorbent, which is called the maximum adsorption space ($W_0$). It is in this space where the adsorbed phase accumulates, in the state of its pure liquid, occupying that part of the adsorption space that minimizes the free energy. The change of free energy associated with the transfer of adsorbate from the bulk of the fluid to the adsorbed phase can be described by the Polyani adsorption potential $\phi$ [1-3] as follows:

$$\phi = -\Delta G = RT \ln \frac{X_s}{X} \qquad (13.1)$$

where $G$ = change in free energy, $R$ = ideal gas constant, $T$ = absolute temperature, $X$ = fluid phase concentration ($X = P$ for gases or $X = C$ for aqueous solution), and $X_s$ = saturation value of $X$.

The value of $\phi$ varies within the adsorption space from a maximum value, $\phi_{max}$, at the surface of the adsorbent to $\phi = 0$ at the interface formed when the adsorbed phase is in equilibrium with a saturated fluid. Points having the same value define an equipotential surface. A series of equipotential surfaces exists within the adsorption space, dividing it into layers as shown in Figure 13.1.

At equilibrium with a given fluid phase concentration, the adsorbed amount is accommodated in the layers of higher adsorption potential. No adsorption space layer can contain adsorbate at equilibrium unless all the layers of higher adsorption potential are full. If the amount adsorbed is increased, e.g., due to higher fluid concentration or lower temperatures, then more layers have to be filled to accommodate the additional amount of adsorbate. At equilibrium with a saturated fluid, the entire maximum adsorption space is filled with adsorbate up to the layer of zero potential.

Therefore, there is a difinite relationship between the adsorbed volume and the adsorption potential of the interface. A plot of the cumulative adsorbed volume versus $\phi$ gives the "characteristic curve" of the system. As far as the van der Waal forces of adsorption are created by the dispersion effect, the characteristic curve of an adsorption system should be independent of temperature. This is because the dispersion effect itself is independent of temperature. For a nonpolar adsorbent like activated carbon, dispersion is likely to be either the only cause of adsorption forces (nonpolar adsorbate) or a major one (polar organic adsorbate).

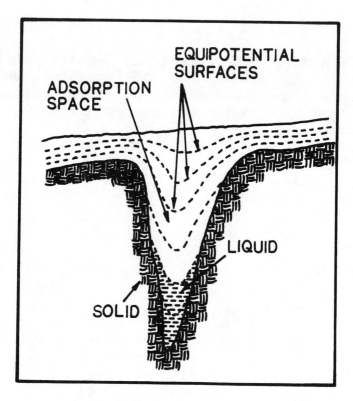

**Figure 13.1.**    Equipotential Surfaces within the Adsorption Space

It has been noticed that a characteristic curve can be expressed mathematically as follows [4,5]:

$$W = W_0 \exp\left[-k\left(\frac{\phi}{b}\right)^m\right] \qquad (13.2)$$

or in a linearized form:

$$\ln W = \ln W_0 - k\left(\frac{\phi}{b}\right)^m \qquad (13.3)$$

where $W_0$ = maximum adsorption space, $k$ = the microporosity of the adsorbent, $\phi$ = characteristic energy constant, $b$ = the affinity coefficient of the adsorbate, $m$ = a constant characterizing the adsorbent. The value of $m$ has been found to be 1 for XAD-4 [6], 1.5 for silica gel [5], and 2 for activated carbon [4]. For adsorption onto carbon, Equation (13.2) reduces to the Dubinin-Radushkevich equation:

$$W = W_0 \exp\left[-k\left(\frac{\phi}{b}\right)^2\right] \qquad (13.4)$$

or

$$\ln W = \ln W_0 - k\left(\frac{\phi}{b}\right)^2 \qquad (13.5)$$

It should be noted that although $m$ characterizes a broad class of adsorbents (*i.e.* $m = 2$ for all the carbons), the other two adsorbent related parameters $W_0$ and $k$, are expressing the micropore volume and structure of the adsorbent and are different for different types of the same adsorbing material. Two bituminous coal-based activated carbons, for example, manufactured under different conditions, may have quite different $W_0$ and $k$ values.

Since $m$, $k$ and $W_0$ are related to the adsorbent only and are independent of the adsorbate, the characteristic curves of all the adsorbates onto the same adsorbent are the same type, and can be condensed to a single characteristic curve by proper normalization. The normalizing factor is the affinity coefficient $b$, of each of the adsorbates.

The affinity of an adsorbate is a relative property, that can be quantified only after an arbitrary $b$ value has been assigned to a reference adsorbate. For convenience, the value $b = 1$, is usually assigned to the adsorbate used to experimentally measure the values of $W_0$, $k$, and eventually $m$, of the particular adsorbent.

Since the adsorption potential theory is based on the assumption that the adsorption forces are generated predominantly by the dispersion effect, the value $b$, should be a function of pertinent properties of the adsorbate. Two such alternative properties that have been applied to predict the affinity factor $b$, of an adsorbate, are the molecular volume and the parachor.

$$b = \frac{V_m}{V_m^0} \qquad (13.6)$$

or

$$b = \frac{P_m}{P_m^0} \qquad (13.7)$$

and

$$P_m = \frac{\theta^{1/4} M}{\rho_a} \qquad (13.8)$$

where $V_m$ = the molecular volume, $P_m$ = the molecular parachor, $\theta$ = the surface tension, $\rho_a$ = the density, and $M$ = the molecular weight of the adsorbate at liquid state. If the surface tension is not available, the parachor can be estimated from the elements, and molecular structure of the adsorbate [7].

In the case of a polar adsorbate, the assumption of dispersion-dominated adsorption, may deviate from reality, as dipole forces may influence adsorption. To account for the effect of those dipole forces, the electronic polarization can be used as the normalizing factor [8]:

$$b = \frac{a}{a^0} \tag{13.9}$$

and

$$a = \frac{(n^2 - 1)M}{(n^2 + 2)} \tag{13.10}$$

where $n$ = the refractive index of the compound at the sodium D wavelength. Polarizability can also be estimated from the elementary composition and molecular structure of the compound [9].

The adsorbed mass of an adsorbate is related to the adsorption space it occupies. As long as the temperature is well below the critical point, the relationship is as follows:

$$q = W\rho_a \tag{13.11}$$

where $q$ = mass adsorbed per unit mass of adsorbent.

*Demonstration of the Method*

The adsorption potential theory was used to describe the adsorption equilibria of six hazardous vapors onto activated carbon BAC and synthetic resin XAD-4. The experiments were conducted at $25^0$C, $50^0$C, and $75^0$C, and the analytical method used was the quartz spring gravimetric technique, described in Chapter IV. The six adsorbates are toluene, $p$-xylene, carbon tetrachloride, benzene, tetrachloroethylene, and trichloroethylene. Their essential properties are shown in Table 4.1. The same equilibrium data has been analyzed in Chapter XII by the Freundlich model resulting in thirty-six different isotherms (Table 12.1). The adsorption isotherms for trichloroethylene for both activated carbon and XAD-4 resin at three different temperatures are shown in Figure 13.2, in the familiar Freundlich log-log plot.

By using the molecular volume as the normalizing factor, the equilibrium data of all the adsorbates and at all three temperatures were condensed into two characteristic lines, one for each of the two adsorbents. This was done by combining Equations (13.1), (13.3), and (13.6), and assuming a hypothetical reference adsorbate with $V_m^0 = 1$; the result is:

$$\ln W = \ln W_0 - K\left(\frac{T_m}{V_m} \log \frac{P_0}{P}\right)^m \tag{13.12}$$

and

$$K = kR^m \tag{13.13}$$

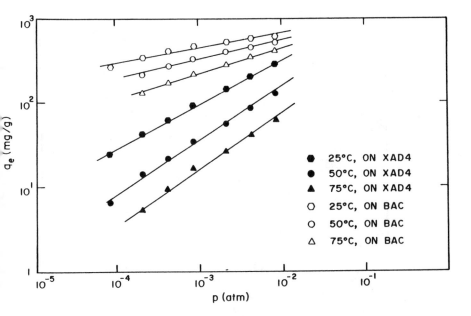

**Figure 13.2.**    Adsorption Isotherms at Different Temperatures

Statistical analysis of all the data revealed that the following values of the coefficients in Equation (13.12):
    for XAD-4:

$$\ln W = \ln 0.645 - 0.168\left(\frac{T_m}{V_m}\ln\frac{P_0}{P}\right) \qquad (13.14)$$

    for BAC:

$$\ln W = \ln 0.451 - 0.004\left(\frac{T_m}{V_m}\ln\frac{P_0}{P}\right)^2 \qquad (13.15)$$

where $W_0 = 0.645$ for XAD-4 and 0.451 for BAC, $k = 8.45 \times 10^{-2}$ for XAD-4 and $1.03 \times 10^{-3}$ for BAC. These equations are shown as the lines in Figures 13.3 and 13.4. The $R^2$ for both the XAD-4 and BAC are equal to 0.97. The parameters obtained by analyzing each adsorbent-adsorbate system separately, are presented in Table 13.1 for comparison.

The good fit of the data to the model demonstrates the validity of this theory. The characteristic curves obtained as described by the $W_0$, $k$, and $m$ values, can be used to predict the capacity of the two adsorbents for other adsorbates at any temperature and concentration. This is done by calculating the quantity $[(T_m/V_m)\log(P_0/P)]$ for an adsorbate of known $V_m$, and $P_0$ at the given conditions $P$ and $T_m$,

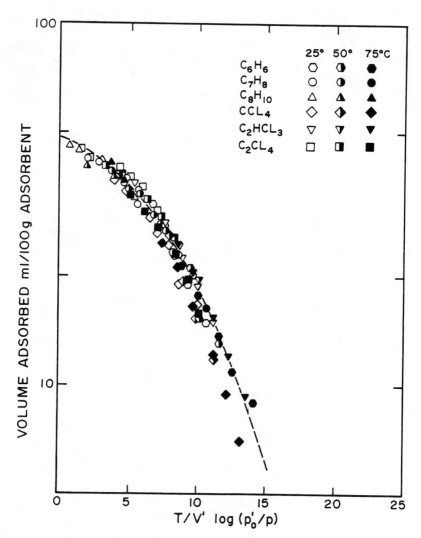

**Figure 13.3.** The Application of the Adsorption Potential Theory for Resin XAD-4

and then use it to obtain (graphically or from Equations (13.14) and (13.15)) the adsorbed volume $W$ per unit mass of adsorbent. Finally, Equation (13.11) can be used to convert $W$ to mass per mass capacity.

Successful application of the adsorption potential theory to describe equilibrium data has been reported by many investigators. Table 13.2 presents the results of published studies and lists the characteristic parameters ($W_0$, $k$, and $m$) of some commercial adsorbents. The corresponding characteristic curves have been plotted in Figure 13.5.

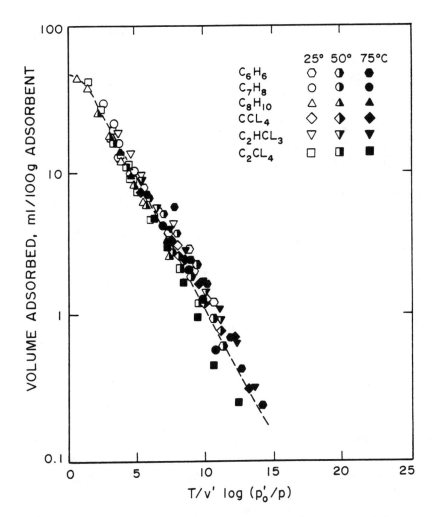

**Figure 13.4.**   The Application of the Adsorption Potential
Theory for Carbon BAC

*Limitations of the Method*

Chemical Groups.   According to the adsorption potential theory,
the adsorption equilibria of any adsorbate onto the same adsorbent
can be described by a single characteristic curve [10,11]. This was
indeed the case with all the hydrocarbons and chlorinated hydrocarbons presented above. It has been observed [12,13], however, that
the adsorption behavior of certain chemical groups of adsorbates may

**Table 13.1.**    Parameters for Correlation Equations

| System | $m$ (−) | $K$ (mole/cal)$^m$ | $W_0$ (m$^3$/g) | $b$ (−) |
|---|---|---|---|---|
| $C_6H_6$/BAC | 2.0 | $2.59 \times 10^{-8}$ | 0.450 | 1.00 |
| $p$-$C_7H_8$/BAC | 2.0 | $1.476 \times 10^{-8}$ | 0.441 | 1.33 |
| $C_8H_{10}$/BAC | 2.0 | $1.342 \times 10^{-8}$ | 0.458 | 1.39 |
| $CCl_4$/BAC | 2.0 | $2.104 \times 10^{-8}$ | 0.430 | 1.11 |
| $C_2HCl_3$/BAC | 2.0 | $2.204 \times 10^{-8}$ | 0.470 | 1.09 |
| $C_2Cl_4$/BAC | 2.0 | $1.389 \times 10^{-8}$ | 0.455 | 1.37 |
| $C_6H_6$/XAD-4 | 1.0 | $3.926 \times 10^{-4}$ | 0.6157 | 1.00 |
| $p$-$C_7H_8$/XAD-4 | 1.0 | $3.791 \times 10^{-4}$ | 0.7895 | 1.04 |
| $C_8H_{10}$/XAD-4 | 1.0 | $3.507 \times 10^{-4}$ | 0.7634 | 1.12 |
| $CCl_4$/XAD-4 | 1.0 | $3.673 \times 10^{-4}$ | 0.6129 | 1.07 |
| $C_2HCl_3$/XAD-4 | 1.0 | $3.938 \times 10^{-4}$ | 0.7804 | 1.00 |
| $C_2Cl_4$/XAD-4 | 1.0 | $3.694 \times 10^{-4}$ | 0.7359 | 1.06 |

deviate from a characteristic curve constructed and followed by other adsorbates. Each of those groups has its own characteristic curve, which is followed by all the adsorbates of the group. Figure 13.6 [14] shows how the adsorption of reduced sulfur compounds onto carbon, defines a characteristic curve, different from that of hydrocarbons. The reasons behind those deviations have not been completely understood yet. Thus, it is recommended that the prediction of adsorption equilibria of a given adsorbate be based on a characteristic curve, constructed by a reference adsorbate(s) of the same chemical group.

Polar Adsorbates. Another cause of deviations from the ideal behavior predicted by the adsorption potential theory, is the polarity of adsorbates. For reasons explained earlier, the ideal application of this theory requires that both the adsorbent and the adsorbate are completely non-polar. Yet, in practice, this theory can well be applied to describe the adsorption of polar adsorbates onto non-polar adsorbents. The errors introduced by the non-idealities of such systems are not intolerable and they can be diminished to below experimental error (±5%), by proper selection of the reference adsorbate and of the normalizing factor. The latter issue, the selection of the best method

**Table 13.2.** Adsorption Parameters for Various Systems

| Description Ref.Solvent | $A_s$ (m²/g) | $\rho_b$ (g/cm³) | $W_0$ (m³/g) | $m$ (-) | $K$ (mol/cal)$^m$ |
|---|---|---|---|---|---|
| **A.** Tsurumi 4GS-S (Coconut Shell) [18] | | | | | |
| Benzene | 1170 | 0.43 | 0.485 | 2.0 | $5.1 \times 10^{-8}$ |
| **B.** Tsurumi HC-8 (Coconut Shell) [18] | | | | | |
| Benzene | 1270 | 0.44 | 0.525 | | $4.7 \times 10^{-8}$ |
| **C.** Takeda SX (Coconut Shell) [18] | | | | | |
| Benzene | 1090 | 0.41 | 0.455 | | $4.2 \times 10^{-8}$ |
| **D.** Hokuetsu Y-20 (Coconut Shell) [18] | | | | | |
| Benzene | 1098 | 0.45 | 0.440 | | $5.1 \times 10^{-8}$ |
| **E.** Fujisawa D-CG (Coconut Shell) [18] | | | | | |
| Benzene | 1240 | 0.43 | 0.495 | | $5.2 \times 10^{-8}$ |
| **F.** Fujisawa A (Coal) [18] | | | | | |
| Benzene | 840 | 0.42 | 0.375 | | $4.6 \times 10^{-8}$ |
| **G.** Kureha G-BAC (Oil Pitch) [18] | | | | | |
| Benzene | 1000 | 0.51 | 0.430 | | $4.6 \times 10^{-8}$ |
| **H.** Columbia JXC (25⁰C, 737 mm) [16,17] | | | | | |
| $n$-Heptane | 1194 | 0.461 | 0.404 | | $1.44 \times 10^{-8}$ |
| $n$-Butanol | | | 0.413 | | |
| MIRK | | | 0.409 | | |
| **I.** Barnebey Cheney (Lot#0993, 23⁰C) [19] | | | | | |
| CCl$_4$ | | | 0.481 | | $1.5 \times 10^{-8}$ |
| **J.** NACAR G-352 (Lot#2905) [19] | | | | | |
| | | 0.347 | 0.700 | | $14.9 \times 10^{-8}$ |
| **K.** Davison PA400 (Silica Gel) [5] | | | | | |
| $n$-Butane | 670 | | 0.318 | 1.5 | $59.2 \times 10^{-6}$ |
| **N,O.** BPL-202654 (Activated Carbon) [14] | | | | | |
| | 1040 | 0.803 | 0.450 | | |
| **L.** Kureha G-BAC (Activated Carbon) [This Study] | | | | | |
| Benzene | 1000 | 0.590 | 0.451 | 2.0 | $2.59 \times 10^{-8}$ |
| **M.** Amberlite XAD-4 (Copolymer) [This Study] | | | | | |
| | 725 | 0.705 | 0.645 | 1.0 | $5.92 \times 10^{-2}$ |

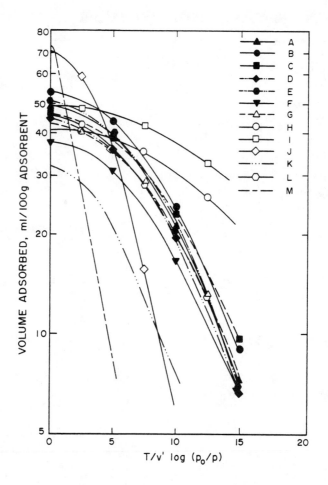

**Figure 13.5.** The Application of Adsorption Potential Theory for Some Systems Described in Table 13.2

to determine affinity coefficients, is important not only for systems involving polar adsorbates, but for any application of this theory in general.

Dubinin and Timofeyev [15] suggested that for non-polar or weakly polar adsorbates, the molar volume or molecular parachor methods may be used to calculate the affinity coefficient. Reucroft *et al.* [8]

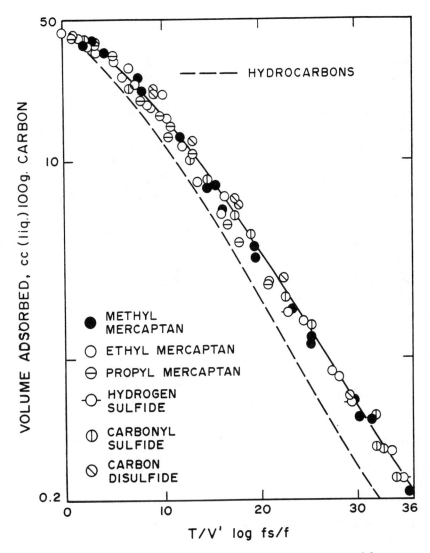

**Figure 13.6.**  The Comparison of Adsorption Potential between Sulfide Compounds and Hydrocarbons

pointed out that for polar adsorbates, the electronic polarization method would be more applicable. They further indicated that for optimum predictive ability, a reference vapor should be chosen, which has similar polarity to the vapor whose isotherm is being predicted. However, Golovoy and Braslaw [16,17] found that the use of a reference compound of similar polarity did not necessarily yield a more accurate prediction, and to a 5% level of significance there is no statistical difference in the accuracy of the results predicted by either of the three methods.

Selection of Reference Adsorbate and Normalizing Method. Recently, a detailed investigation focused on these two questions was conducted by Noll et al. [6]. They studied experimentally the adsorption of ten organic vapors, ranging from non-polar to strongly polar onto activated carbon BAC, and evaluated the predictions of the adsorption potential theory offered by alternative reference adsorbates and by each of the three methods for affinity estimation. The dipole moment and the other parameters of these adsorbates are listed in Table 13.3. According to the values of the dipole moment, these organic compounds can be divided into two groups: (a) non-polar and weakly polar materials including benzene, carbon tetrachloride, p-xylene, tetrachloroethylene, toluene and trichloroethylene, and (b) medium and strongly polar materials, including n-butanol, ethyl acetate, pyridine and acetone.

Two of those compounds, non-polar (benzene), and polar (ethyl acetate), were used as reference adsorbates independently, yielding two characteristic curves expressed according to Equation (13.3) as follows:

$$\ln W = \ln 0.481 - 2.22 \times 10^{-8} \phi^2 \quad \text{for Benzene} \qquad (13.16)$$

$$\ln W = \ln 0.495 - 2.19 \times 10^{-8} \phi^2 \quad \text{for Ethyl Acetate} \qquad (13.17)$$

The correlation coefficient for Equations (13.16) and (13.17) exceeded 0.99. The parameter $W_0$, representing the active pore volume of the carbon, had an average value of 0.49 cm$^3$/g for the two referenced adsorbates. This value is similar to the values reported by Urano et al. [18]. They reported that the values of $W_0$ for activated carbon columns are from 0.430 to 0.525 cm$^3$/g. With the values of $W_0$ and $k$ determined above and the theoretical affinity coefficients shown in Table 13.3, Equation (13.4) can be used to predict the isotherms of the organic vapors used in this study.

In order to effectively make comparisons and draw conclusions from the experimental results, they grouped the data obtained into two systems: similar systems (non-polar adsorbate/non-polar reference vapor or polar adsorbate/polar reference vapor), and non-similar systems (non-polar adsorbate/polar reference vapor or polar adsorbate/non-polar reference vapor). The isotherm data of similar and non-similar systems predicted by the three methods (molar volume, molecular parachor and electronic polarization), are listed in Tables 13.4 and 13.5, respectively. The relevant experimental isotherm data for both

Table 13.3.  Properties of Adsorbates at 25°C

| Adsorbate | Dipole Moment Debye | Molar Volume $V_m$ | Molecular Parachor $P_m$ | Electronic Polarization $a$ |
|---|---|---|---|---|
| Benzene | 0 | 98.2 (1/0.912) | 206 (1/0.954) | 26.1 |
| Carbon Tetrachloride | 0 | 86.5 (1.08/0.987) | 219 (1.06/1.01) | 26.4 |
| p-Xylene | 0 | 123 (1.39/1.26) | 285 (1.39/1.32) | 36.0 |
| Tetrachloro-ethylene | 0 | 102 (1.15/1.04) | 243 (1.18/1.13) | 30.3 |
| Toluene | 0.36 | 106 (1.19/1.08) | 244 (1.19/1.13) | 30.9 |
| Trichloro-ethylene | 0.85 | 90.0 (1.01/0.92) | | 25.3 |
| n-Butanol | 1.66 | 91.6 (1.01/0.937) | 201 (0.976/0.931) | 22.1 |
| Ethyl Acetate | 1.80 | 97.8 (1.10/1.00) | 216 (1.05/1.00) | 22.1 |
| Pyridine | 2.19 | 80.5 (0.903/0.823) | 200 (0.973/0.926) | 24.0 |
| Acetone | 2.88 | 73.5 (0.825/0.752) | 162 (0.790/0.750) | 16.1 |

* Estimated value from the dipole moment data = Trichloroethylene(1.01), Trifluoroethylene(1.40), Trifluoroethane(1.65).
† Numbers in parentheses = the values of affinity coefficient based on Benzene and Ethyl Acetate as a reference adsorbate, respectively.

systems are shown in the last column of Tables 13.4 and 13.5. The numbers in parentheses alongside the calculated isotherm values represent the relative percent deviation from the experimental values.

As seen from Tables 13.4 and 13.5, the total mean percent deviation of the similar systems (about 3.6%) is smaller than that of the

**Table 13.4.** Calculated and Experimental Values of Equilibrium Adsorption Capacity (mg adsorbate/g carbon) at 25°C, 760 mm Hg with Similar Polarity Reference Vapors

| $C_v$(ppm) | M.V.[†] | M.P. | E.P. | Exp. |
|---|---|---|---|---|
| Carbon Tetrachloride (Adsorbate) Benzene (Reference) | | | | |
| 191 | 388.5(4.7)* | 379.6(2.3) | 350.8(5.4) | 371.0 |
| 386 | 445.0(2.4) | 436.8(0.5) | 410.1(5.6) | 434.6 |
| 1909 | 573.0(1.6) | 567.3(0.6) | 548.5(2.7) | 563.9 |
| 3842 | 624.2(3.9) | 619.8(1.9) | 605.2(0.5) | 608.4 |
| | Av.Er.(3.2) | Av.Er.(1.3) | Av.Er.(3.6) | |
| p-Xylene (Adsorbate) Benzene (Reference) | | | | |
| 150 | 346.9(1.4) | 347.1(1.4) | 346.3(1.6) | 351.9 |
| 600 | 381.4(0.7) | 381.5(0.7) | 381.1(0.6) | 378.8 |
| 1500 | 398.0(0.5) | 398.0(0.5) | 347.8(0.4) | 396.2 |
| 6000 | 412.0(0.1) | 412.0(0.1) | 412.0(0.1) | 411.5 |
| | Av.Er.(0.7) | Av.Er.(0.7) | Av.Er.(0.7) | |
| Tetrachloroethylene (Adsorbate) Benzene (Reference) | | | | |
| 181 | 562.8(12.5) | 573.3(10.7) | 566.6(11.4) | 643.2 |
| 733 | 659.8(7.7) | 666.0(6.8) | 662.0(7.4) | 714.9 |
| 1810 | 711.1(4.1) | 714.8(3.6) | 712.4(3.4) | 741.5 |
| 7275 | 764.0(1.6) | 764.8(1.5) | 764.3(1.5) | 776.3 |
| | Av.Er.(6.5) | Av.Er.(5.7) | Av.Er.(6.2) | |
| n-Butanol (Adsorbate) Ethyl Acetate (Reference) | | | | |
| 201 | 295.2(4.2) | 294.0(4.6) | 306.4(0.6) | 308.2 |
| 806 | 352.9(2.6) | 352.3(2.4) | 358.4(4.2) | 344.0 |
| 4034 | 394.1(6.7) | 394.0(6.7) | 394.8(6.9) | 369.4 |
| 8064 | 400.1(7.3) | 400.1(7.3) | 400.2(7.3) | 372.9 |
| | Av.Er.(5.2) | Av.Er.(5.3) | Av.Er.(4.8) | |
| Pyridine (Adsorbate) Ethyl Acetate (Reference) | | | | |
| 229 | 267.6(15.6) | 303.3(4.3) | 345.0(8.9) | 316.9 |
| 917 | 359.8(8.7) | 383.3(2.7) | 409.0(3.8) | 393.9 |
| 4587 | 447.2(0.7) | 455.1(1.1) | 463.4(2.9) | 450.3 |
| 22936 | 482.2(0.4) | 483.0(0.5) | 483.8(0.7) | 480.4 |
| | Av.Er.(6.4) | Av.Er.(2.2) | Av.Er.(4.1) | |

*(continued)*

**Table 13.4.** Calculated and Experimental Values of Equilibrium Adsorption Capacity (mg adsorbate/g carbon) *(continued)*

| $C_v$(ppm) | M.V.[†] | M.P. | E.P. | Exp. |
|---|---|---|---|---|
| Acetone (Adsorbate) Ethyl Acetate (Reference) | | | | |
| 251 | 82.9(0.9) | 82.2(0.0) | 25.0(8.8) | 82.2 |
| 1004 | 844.0(1.4) | 143.3(0.9) | 135.1(4.9) | 143.0 |
| 5024 | 234.2(2.2) | 233.5(1.9) | 226.6(1.1) | 229.1 |
| 25120 | 324.6(3.5) | 324.2(3.4) | 320.7(2.3) | 313.6 |
|  | Av.Er.(2.0) | Av.Er.(3.0) | Av.Er.(4.3) | |

| | | | |
|---|---|---|---|
| Mean Er. (%) | (4.0) | (2.8) | (3.9) |
| Max. Er. (%) | (15.6) | (10.7) | (11.9) |

* Number in parentheses = relative percent deviation from experimental values.

† Capacity $Q_e$ from different methods: M.V.=Molar Volume, M.P.= Molecular Parachor, E.P.=Electronic Polarization, Exp.=Experimental.

non-similar systems (about 7.2%). The maximum error of the similar system (about 15.6%) is much lower than that of the non-similar systems (about 55.8%). Thus, the results above demonstrated that for optimum predictive ability, the chosen reference vapor should have similar polarity to the vapor whose adsorption isotherm is being predicted.

Although a few groups, for example, tetrachloroethylene/molecular parachor, *n*-butanol/molar volume and *n*-butanol/molecular parachor (see Tables 13.4 and 13.5), have the opposite condition in which the mean deviation of the similar system is greater than that of the non-similar system, most of the groups are consistent with the conclusion above. Golovoy and Braslaw [16] indicated that to a 5% level of significance, the accuracy of the predicted results for low, medium, and high polarity groups does not depend on the polarity of the reference solvent. But actually, if their data had been divided into two systems, similar and non-similar systems, the same conclusion as above would

**Table 13.5.**   Calculated and Experimental Values of Equilibrium
Adsorption Capacity (mg adsorbate/g carbon) at $25^0$C,
760 mm Hg with Different Polarity Reference Vapors

| $C_0$(ppm) | M.V.[†] | M.P. | E.P. | Exp. |
|---|---|---|---|---|
| Benzene (Adsorbate) Ethyl Acetate (Reference) | | | | |
| 210 | 182.2(8.4) | 196.3(1.3) | 258.6(30.1) | 198.8 |
| 425 | 218.3(7.9) | 231.6(2.3) | 288.0(21.5) | 237.0 |
| 850 | 255.4(6.5) | 267.3(2.2) | 316.3(15.7) | 273.3 |
| 2100 | 304.1(5.3) | 313.6(2.3) | 351.0(9.3) | 321.1 |
| 4232 | 339.8(2.3) | 347.0(0.2) | 375.0(7.8) | 347.8 |
| 8438 | 371.5(0.2) | 376.5(1.5) | 395.4(6.6) | 370.8 |
| 20630 | 404.7(5.3) | 407.1(6.0) | 416.1(8.3) | 384.2 |
| | Av.Er.(5.1) | Av.Er.(3.2) | Av.Er.(14.2) | |
| Carbon Tetrachloride (Adsorbate) Ethyl Acetate (Reference) | | | | |
| 191 | 352.4(5.0) | 367.7(0.9) | 455.3(22.7) | 371.0 |
| 386 | 414.0(4.7) | 428.3(1.4) | 508.2(16.9) | 434.6 |
| 1909 | 558.7(0.9) | 568.4(0.9) | 623.5(10.6) | 563.9 |
| 3842 | 618.3(1.6) | 626.3(2.9) | 668.1(9.8) | 608.4 |
| | Av.Er.(3.1) | Av.Er.(1.5) | Av.Er.(15.0) | |
| p-Xylene (Adsorbate) Ethyl Acetate (Reference) | | | | |
| 150 | 385.1(9.4) | 350.5(0.4) | 375.5(6.7) | 351.9 |
| 600 | 431.2(13.8) | 389.2(2.7) | 401.7(6.0) | 378.8 |
| 1500 | 453.7(14.5) | 407.9(3.0) | 414.1(4.5) | 396.2 |
| 6000 | 472.9(14.9) | 423.8(3.0) | 424.5(3.2) | 411.5 |
| | Av.Er.(14.9) | Av.Er.(2.3) | Av.Er.(5.1) | |
| Tetrachloroethylene (Adsorbate) Ethyl Acetate (Reference) | | | | |
| 181 | 543.5(15.5) | 574.0(10.8) | 640.3(0.5) | 643.2 |
| 733 | 657.3(8.1) | 675.9(5.5) | 714.7(0.0) | 714.9 |
| 1810 | 718.8(3.1) | 729.9(1.6) | 752.7(1.5) | 741.5 |
| 7275 | 783.2(0.9) | 785.3(1.2) | 791.0(1.9) | 776.3 |
| | Av.Er.(6.9) | Av.Er.(4.8) | Av.Er.(9.7) | |

*(continued)*

**Table 13.5.**  Calculated and Experimental Values of Equilibrium
Adsorption Capacity (mg adsorbate/g carbon) *(continued)*

| $C_0$(ppm) | M.V.[†] | M.P. | E.P. | Exp. |
|---|---|---|---|---|
| *n*-Butanol (Adsorbate) Benzene (Reference) | | | | |
| 201 | 301.3(2.2) | 292.6(5.1) | 265.8(13.8) | 308.2 |
| 806 | 350.0(1.7) | 345.8(0.5) | 332.3(3.4) | 344.0 |
| 4034 | 383.9(3.9) | 383.3(3.8) | 381.4(3.2) | 369.4 |
| 8064 | 388.9(4.3) | 388.8(4.3) | 388.7(4.2) | 372.9 |
| | Av.Er.(3.0) | Av.Er.(3.4) | Av.Er.(6.2) | |
| Ethyl Acetate (Adsorbate) Benzene (Reference) | | | | |
| 189 | 233.6(11.6) | 220.0(5.1) | 152.6(27.1) | 209.4 |
| 381 | 265.2(6.8) | 212.8(1.8) | 189.1(23.9) | 248.4 |
| 755 | 295.7(1.8) | 284.9(1.9) | 227.3(21.7) | 290.4 |
| 1892 | 335.3(2.1) | 327.0(0.5) | 281.0(14.5) | 328.5 |
| 3779 | 362.5(1.2) | 356.3(2.9) | 320.7(12.6) | 366.9 |
| 7554 | 386.4(0.6) | 382.1(1.7) | 357.2(8.1) | 388.8 |
| 18895 | 411.5(0.4) | 409.5(0.0) | 397.3(3.0) | 409.7 |
| | Av.Er.(3.5) | Av.Er.(2.0) | Av.Er.(15.4) | |
| Pyridine (Adsorbate) Benzene (Reference) | | | | |
| 229 | 285.7(9.8) | 306.3(3.3) | 290.1(8.5) | 316.9 |
| 917 | 366.7(6.9) | 379.8(3.6) | 369.5(6.2) | 393.9 |
| 4587 | 440.3(2.2) | 444.6(1.3) | 441.3(2.0) | 450.3 |
| 22936 | 469.1(2.4) | 469.6(2.2) | 469.2(2.3) | 480.4 |
| | Av.Er.(5.3) | Av.Er.(2.6) | Av.Er.(4.8) | |
| Acetone (Adsorbate) Benzene (Reference) | | | | |
| 251 | 102.9(24.6) | 91.4(10.7) | 36.5(55.8) | 82.2 |
| 1004 | 163.9(14.6) | 151.8(6.2) | 84.0(41.3) | 143.0 |
| 5024 | 241.7(7.7) | 237.3(3.4) | 175.1(23.6) | 229.1 |
| 25120 | 324.8(3.6) | 320.2(2.3) | 286.7(8.6) | 313.6 |
| | Av.Er.(12.6) | Av.Er.(5.7) | Av.Er.(32.3) | |
| Mean Er. (%) | (6.2) | (2.9) | (12.4) | |
| Max. Er. (%) | (24.6) | (10.8) | (55.8) | |

have been found. For example, the mean errors of the similar system of their publication are 6.5% and 6.0%, while the mean errors of the nonsimilar system are 11.5% and 14.0%.

For the second question, the data in Tables 13.4 and 13.5 show different results for the two systems. For similar systems, the mean error is 4.0% for the molar volume method, 2.8% for the molecular parachor method, and 3.9% for the electronic polarization method. It is quite evident that there is essentially no difference in the accuracy of the results calculated by the three methods for similar systems. A possible exception to this conclusion involves the medium and strongly polar group of the similar systems. For this group, the use of polarity as the normalizing factor, although it did not always offer the closest predictions, was the only method that consistently yielded predictions with less than 5% error from the experimental measurements. More data are required to generalize this observation. For the nonsimilar systems, however, there was a remarkable difference in the accuracy of the three methods. In this case, the maximum mean error (about 12.4%) is about four times higher than the minimum mean error (about 2.9%).

Since the predictions for the similar systems show similar accuracy for the three methods, the molar volume method is generally recommended; this is because the physical parameters included in the other two methods are not always available in the literature. The major volume method requires only the determination of the parameters $M$ and $\rho_a$. $M$ is a constant and changes very slightly with variation of temperature. So the molar volume method is quite readily used to predict the adsorption isotherms at different temperatures. In order to test the conclusion above, the isotherms of two weakly polar organic compounds, toluene and trichloroethylene, were predicted at $25^0$C, $40^0$C, and $60^0$C by using Equation (13.4) and the molar volume, based on benzene as a reference vapor. The calculated and experimental results are listed in Table 13.6.

As seen from Table 13.6, the affinity coefficients at $25^0$C, $40^0$C, and $60^0$C are almost the same for the two adsorbates, and the average deviation between the calculated and experimental values at the three temperatures is about 4.4%. Therefore, the prediction of the adsorption isotherm at different temperatures by using the molar volume method is satisfactory.

*CONCLUSIONS*

In summary, some conclusions can be obtained from this study:

**Table 13.6.** Calculated and Experimental Values of Equilibrium Adsorption Capacity (mg adsorbate/g carbon) at $25^0$C, $40^0$C, and $60^0$C. Reference Adsorbate: Benzene.

| Adsorbate | $T(^0C)$ | $a$ | $C_0$(ppm) | M.V. | Exp. |
|---|---|---|---|---|---|
| Toluene | 25 | 1.19 | 350 | 316.7(8.6) | 346.5 |
| | | | 700 | 341.6(7.0) | 367.5 |
| | | | 1700 | 369.7(4.4) | 386.7 |
| | | | 3550 | 388.9(2.6) | 399.3 |
| | | | | Av.Er.(5.7) | |
| | 40 | 1.20 | 170 | 247.8(8.5) | 270.7 |
| | | | 700 | 305.8(4.4) | 320.0 |
| | | | 3550 | 363.7(3.3) | 364.9 |
| | | | 17100 | 401.8(2.0) | 394.0 |
| | | | | Av.Er.(4.6) | |
| | 60 | 1.21 | 170 | 195.3(0.7) | 193.4 |
| | | | 700 | 255.5(1.3) | 258.8 |
| | | | 3550 | 324.0(0.4) | 325.4 |
| | | | 17100 | 377.1(2.4) | 386.2 |
| | | | | Av.Er.(1.2) | |
| Trichloro-ethylene | 25 | 1.01 | 206 | 360.6(10.2) | 401.5 |
| | | | 417 | 416.4(8.6) | 455.5 |
| | | | 2062 | 540.9(4.0) | 563.6 |
| | | | 4155 | 589.6(2.7) | 605.9 |
| | | | | Av.Er.(6.4) | |
| | 40 | 1.01 | 417 | 338.4(6.1) | 360.4 |
| | | | 834 | 395.3(7.0) | 424.9 |
| | | | 2062 | 471.3(6.9) | 506.3 |
| | | | 4155 | 527.8(4.5) | 552.9 |
| | | | | Av.Er.(6.1) | |
| | 60 | 1.02 | 206 | 201.1(0.4) | 201.9 |
| | | | 834 | 308.8(2.0) | 310.0 |
| | | | 4155 | 440.9(4.4) | 461.4 |
| | | | 20245 | 571.2(3.4) | 591.2 |
| | | | | Av.Er.(2.6) | |

1. For optimum prediction of equilibrium values by using the potential theory, the reference vapor should have similar polarity to the vapor whose adsorption is being predicted.
2. After making an appropriate choice for the reference adsorbate, there is no obvious difference in the accuracy of the isotherm prediction for the three methods. For polar adsorbates, however, the polarity method may be more consistent.
3. Because of its simplicity and requirements for easily obtainable physical parameters, the molar volume method is recommended to predict isotherms at different temperatures.

The adsorption potential theory can be expanded to predict the equilibria of multicomponent adsorption systems. The same adsorption space occupied by an adsorbate, or mixture of adsorbates, corresponds always to the same value of normalized adsorption potential, as defined by the characteristic curve. This realization is the key for the analysis of multicomponent adsorption, since it relates to the equilibrium concentrations of the components of a mixture to each other, and to those of pure component systems. Its extension to more complicated systems is straightforward.

*REFERENCES*

1. Manes, M. *The Polanyi Adsorption Potential Theory and Its Applications to Adsorption from Water Solution onto Activated Carbon.* in *Activated Carbon Adsorption.* Vol. 1, Suffet, I.H., and M.J. McGuire ed., Ann Arbor Science Publishers Inc., Ann Arbor, MI.
2. Manes, M., and L.J.E. Wafer, *J. Phys. Chem.*, **73**, 584 (1969).
3. Schenz, T.W., and M. Manes, *J. Phys. Chem.*, **79**, 604 (1975).
4. Dubinin, M.M., and L.V. Radushkevich, *Dokl. Akad. nauk USSR*, **55**, 331 (1947).
5. Al-Sahhaf, T.A., *Ind. Eng. Chem. Process Des. Dev.*, **20**, 658 (1981).
6. Noll, K.E., D. Wang, and T. Shen, *Carbon*, in press, (1988).
7. Reid, R.C., T.M. Prausnitz, and T.K. Sherwood, *The Properties of Gases and Liquid*, McGraw-Hill, New York (1977).
8. Reucroft, P.J., W.H. Simpson, and L.A. Jonas, *J. Phys. Chem.*, **75**, 3526 (1971).
9. Perry, R.H., and C.H. Chilton, *Chemical Engineering Handbook*, 6th ed., McGraw-Hill Co., New York (1986).
10. Rosene, M.R., and M. Manes, *J. Phys. Chem.*, **80**, 953 (1976).
11. Rosene, M.R., and M. Manes, *J. Phys. Chem.*, **81**, 1646 (1977).
12. Wohleber, D.A., and M. Manes, *J. Phys. Chem.*, **75**, 4 & 20 (1970).
13. Kuennen, R., K. Van Dyke, J.C. Crittenden, and D. Hand, *Prediction of Multicomponent Fixed-Bed Adsorber Performance Using Mass Transfer and Thermodynamic Models.* AWWA. Meeting (1988).
14. Grand, R.J. *AIChE. J.*, **8**, 403 (1962).
15. Dubinin, M.M., and D.P. Timofeyev, *Zh. Fiz. Khim.*, **22**, 113 (1948).
16. Golovoy, A., and J. Braslaw, *J. APCA.*, **31**, 861 (1981).
17. Golovoy, A., and J. Braslaw, *Envir. Prog.*, **1**, 89 (1982).
18. Urano, K., S. Omori, and E. Yamamoto, *Envir. Sci. & Tech.*, **16**, 10 (1982).

# CHAPTER XIV

## EVALUATION OF HUMIDITY EFFECTS FOR GAS ADSORPTION SYSTEMS

Contributing Author: D.H. Wang

## INTRODUCTION

Adsorption is a process that is used for the purification of both aqueous and gaseous streams. In general, the sorptive removal of a given pollutant from a gaseous stream is much more efficient than its removal from water. The presence, however, of high humidity in a gaseous stream can interfere with the vapor adsorption process and diminish its efficiency to levels far below the efficiency of the aqueous adsorption. The loss of capacity due to high humidity is especially severe for adsorbates of low water solubility. For certain applications, the difference in efficiency between air and water adsorption, as well as the humidity interference, are of practical importance; the most typical example is the removal of volatile organic compounds from contaminated groundwater and soil.

The volatile organic compounds (VOC's) are a class of adsorbable pollutants that can contaminate all three constituents of the environment; the air, the water, and the soil. Those compounds are slightly soluble in water, but many of them are toxic and carcinogenic. Their presence in potable aquifiers, even at trace concentrations, poses a serious health hazard. Two processes are commonly used for their removal; air-stripping and carbon adsorption.

Stripping merely transfers those pollutants from the contaminated water or soil to the air. Due to increased concern about air quality, air-stripping alone, is not an acceptable treatment method for many environmental conditions. It can be used, however, as a pretreatment step to volatilize the VOC's so that they can be subsequently adsorbed. Comparative studies [1,2] concluded that it is cheaper to treat a VOC-contaminated stream by air stripping and subsequent vapor adsorption than by direct application of the adsorption process to the water stream. In the case of VOC contaminated soil, adsorption is technically impossible without prior air stripping of the soil [3]. The adsorption of VOC's emitted from a stripping operation, however, can be hindered because the off-gas stream has a high humidity level, approaching saturation, and the pollutants to be removed are slightly soluble in water.

The effect of humidity on organic vapor adsorption is analyzed quantitatively in the present chapter. An evaluation of all the factors involved allows the identification and selection of those conditions under which the humidity interference and the associated loss of capacity are minimized. Although the adsorbates and conditions used for study and demonstration of the methods presented are related to the currently important issue of VOC control, all principles, methods, and conclusions are of general applicability and valid for any case of vapor adsorption from a humid stream.

*Sorptive Behavior of Water*

Water has a complicated sorptive behavior as revealed by the shape of the adsorption isotherm shown in Figure 14.1 [4]. This is an *S*-shaped isotherm, indicating that more than one mechanism is responsible for the adsorption of water vapor onto carbon. The first part of this isotherm, extending from zero relative humidity (R.H.) to R.H.=40–50%, shows a weak monolayer adsorption. This is expected since the strongly polar water vapor is unlikely to find a significant number of hydrophilic adsorption sites on the surface of a nonpolar adsorbent like activated carbon. At higher R.H. values ($> 40 - 50\%$), however, a sharp increase of the adsorption capacity is noticed, indicating capillary condensation inside the pores of the carbon.

The exact point at which capillary condensation starts taking place and the extent to which it will proceed at any R.H. value, as expressed by the middle part of the isotherm, depends on the pore structure and size distribution of the particular carbon. The condensed water starts filling the smaller pores first and subsequently occupies the larger ones, to the extent permitted by the R.H., the surface tension, and the pore radius. Figures 14.2 to 14.5 show experimentally determined water isotherms for various types of activated carbon [5].

*Okazaki's Model*

According to a model proposed by Okazaki *et al.* [6], the partial filling of the pores by condensed water involves three distinct means by which the organic adsorbate can accumulate: (1) a dry part, where ordinary vapor adsorption can take place on the pore surface, (2) a condensed water part, where the vapor can be dissolved in the water, and (3) a wet part, where the dissolved organic can adsorb by the

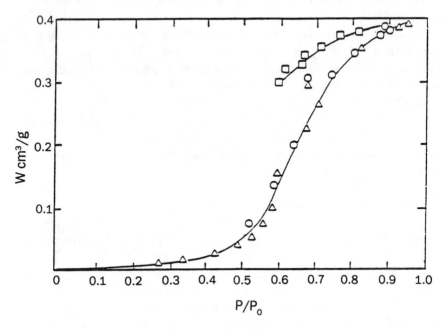

**Figure 14.1.**    Water Vapor Isotherm

mechnisms of adsorption from the aqueous solution. This is shown schematically in Figure 14.6. The total adsorption capacity for the organic adsorbate is given as the summation of the amount adsorbed in each of the three areas. An approach of estimating each of these quantities is described below.

Estimation of the Wet and Dry Parts of the Adsorbent.   To estimate the amount of organic adsorbate adsorbed on the dry surface, dissolved in the condensed water, or adsorbed on the wet surface, one must know at any given R.H. how much water is condensed, how much of the adsorbent surface is wetted, and how much surface remains dry.

Okazaki proposed that the capillary condensation of water in the pores of carbon is independent of any "water insoluble" organic adsorbate present. This means that the volume of the condensate is a function of the R.H. only, and can be estimated readily at any R.H. from a water isotherm like the one in Figure 14.1. Recent findings [7] contradict this assumption, as will be discussed later. For the moment, we will accept that Okazaki's assumption introduces a tolerable error if restricted to systems with extensive condensation (high R.H.) and sufficiently low organic adsorbate concentration.

Having estimated the volume of the condensate, a relationship be-

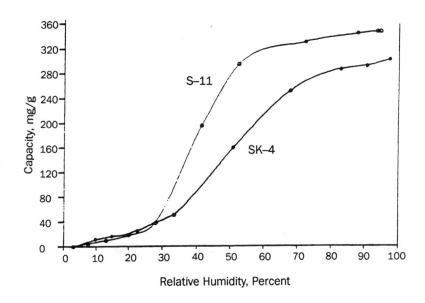

**Figure 14.2.**    Water Adsorption Isotherms for SK-4 and S-11

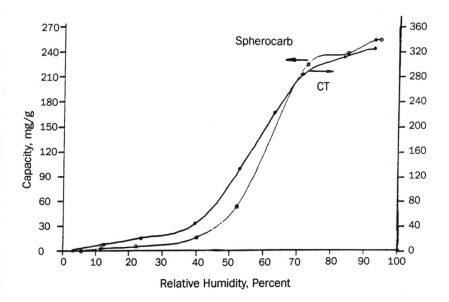

**Figure 14.3.**    Water Adsorption Isotherms for Spherocarb and CT

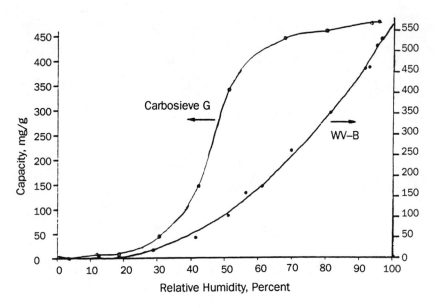

**Figure 14.4.**     Water Adsorption Isotherms for Carbosieve G and
WV-B

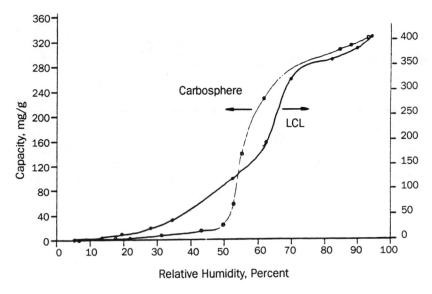

**Figure 14.5.**     Water Adsorption Isotherms for LCL and
Carbosphere

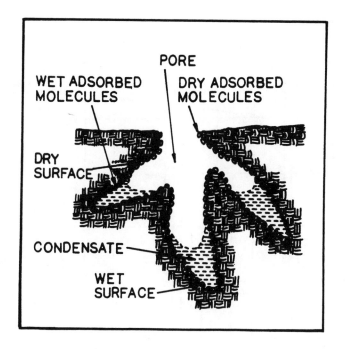

**Figure 14.6.** Partial Filling of Condensed Water in the Pores

tween the pore volume and the pore surface is needed to determine the wet and dry fractions of the pore surfanal business relationship, which is different for each carbon. This can be derived from the pore size-pore volume distribution (Figure 14.7). The latter can be obtained experimentally [8] and is usually supplied by the carbon manufacturer. This derivation is based on Equation (14.1):

$$S = 2 \sum_{r_p=0}^{r_p} \frac{\Delta V}{\overline{r}_p} \tag{14.1}$$

$$\Delta V = V_{r_p + \Delta r_p} - V_{r_p} \tag{14.2}$$

$$\overline{r}_p = \frac{r_p + (r_p + \Delta r_p)}{2} \tag{14.3}$$

where $S$ = cumulative surface of the pores having radius smaller or equal to $r_p$, and $V$ = cumulative volume of the pores having radius smaller or equal to $r_p$.

The relationship between $S$ and $V$, corresponding to the same $r_p$, is called pore area-pore volume distribution. An example of this distribution is presented in graphical form in Figure 14.8. It refers to

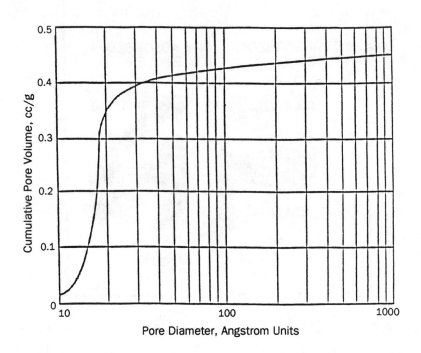

**Figure 14.7.**   Pore Size and Pore Volume Distribution of BPL GAC

BPL activated carbon (Calgon) and has been derived from the pore size-pore volume distribution of Figure 14.7. A graph, like the one in Figure 14.8, provides the wetted and dry pore surface area of a particular carbon at any known volume of condensate. The total pore surface area is provided by both the graphs appearing in Figures 14.7 and 14.8. By dividing the wet and dry pore area by the total pore area, the wet and dry fractions of the pore surface are obtained.

Organic Adsorbed on the Dry Surface. The contribution of the dry surface to the total adsorption capacity for the organic is given by Equation (14.4):

$$Q_1 = Q_1^0 Z_d \qquad (14.4)$$

where $Z_d$ = the dry fraction of the pore surface and $Q_1^0$ = the local equilibrium solid concentration at the dry part of the adsorbent. It should be noted that $Q_1^0$ is expressed in units of mass adsorbed per unit mass of "dry" adsorbent, while $Q_1$ in mass adsorbed per unit mass of "total" (both wet and dry) adsorbent. $Q_1^0$ is the adsorbent capacity at dry conditions and its value can be obtained by any vapor

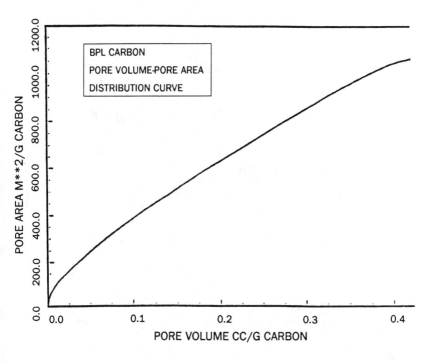

**Figure 14.8.**    Pore Volume and Pore Surface Area of BPL GAC

adsorption isothermal model.

It is, however, highly desirable that the value of $Q_1^0$ be calculated by the adsorption potential theory to permit evaluation of the adsorption capacity at different temperatures. This is because temperature affects both adsorption and condensation. As temperature rises, the dry adsorption capacity is reduced but the relative humidity of the gas stream is also reduced resulting in less capacity loss due to water vapor interference. As the psychrometric chart in Figure 14.9 indicates, a 10–15°C temperature increase reduces a 100% R.H. to levels (40-50%), at which no significant effect on adsorption is expected. The result of this conflicting effect of temperature is the existence of an optimum temperature, at which the overall capacity is maximized. This makes temperature control a major aspect of process optimiza-

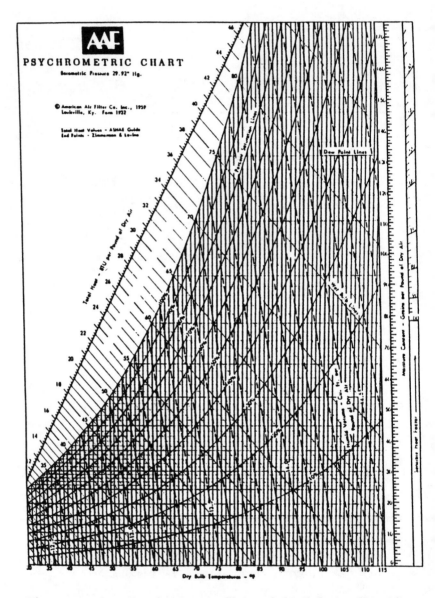

**Figure 14.9.**    Psychrometric Chart of the Relative Humidity

tion. It is, therefore, imperative that the effect of temperature be
included in modeling of the process. The only adsorption equilibria
model that predicts the adsorption capacity at different temperatures,
is the adsorption potential theory.

Equation (14.4) assumes that there is no interaction between the
organic adsorbate and the water adsorbed on the dry part of the

adsorbent. This is reasonable, since the hydrophobic nature of the dry carbon surface excludes any significant adsorption of water. The assumption above has been confirmed in practice by the numerous applications of the adsorption process under humid conditions (but not humid enough for capillary condensation to occur (R.H.<40%)) without any apparent loss of capacity.

Organic Dissolved in the Condensate. The estimation of the amount of water condensed was discussed above. The amount $Q_2$ of the organic dissolved in units of mass of organic per unit mass of carbon is:

$$Q_2 = V_w C \qquad (14.5)$$

where $V_w$ = volume of water condensed per unit mass of carbon and $C$ = concentration of the organic dissolved in condensate.

Since the condensate is in equilibrium with the organic vapor in the gaseous stream, the value of $C$ is calculated from the Henry's law constant, correcting for the effect of curvature of the concave meniscus. This curvature effect correction stems from the fact that the mechanical equilibrium of a curved interface has to be justified in addition to chemical equilibrium [9]. Following the analysis presented by Defay et al. [9], the equilibrium between the vapor phase of an organic and its capillary water solution is expressed by Equation (14.6) [10]:

$$\ln \frac{P}{P_0} = \frac{-2\theta V' \cos a_c}{r_p R T} \qquad (14.6)$$

where $P$ = the vapor pressure of the organic in the stream and in equilibrium with $C$ at the curved capillary interface, $P_0$ = the idealized vapor pressure of the organic in equilibrium with $C$ at a flat interface, $\theta$ = surface tension of the condensate, $V'$ = molar volume of the solute, and $a_c$ = contact angle of the meniscus.

The value of $P_0$ is given by Henry's law:

$$P_0 = HC \qquad (14.7)$$

As far as the organic compound is "insoluble" in water, the contact angle, $a_c$, and the surface tension, $\theta$, may be assumed to be equal to those of pure water [6]. Considering a pure water condensate and applying Equation (14.6) on the water vapor:

$$\ln \frac{P_w}{P_{w0}} = \ln(R.H.) = \frac{-2\theta V'_w \cos a_c}{r_p R T} \qquad (14.8)$$

where the subscript $w$ refers to water. Dividing Equation (14.7) by Equation (14.8) and substituting $P_0$ from Equation (14.7):

$$C = \frac{P}{H}(R.H.)^{-V'/V'_w} \tag{14.9}$$

This equation indicates that the solute concentration in the case of a curved interface is always higher than the concentration predicted by the Henry's law for the case of flat interface.

Organic Adsorbed on the Wet Surface. Since the liquid phase concentration, $C$, is fixed by the vapor-condensate equilibrium, the solid phase concentration on the wet surface is readily determined from the aqueous adsorption isotherm. Although any valid isothermal model can be used to calculate this solid concentration, the use of the adsorption potential theory is again, as in the case of $Q_1^0$, essential to allow capacity evaluations at different temperatures.

The adsorption potential of an aqueous adsorbate is given as:

$$\epsilon = RT \ln \frac{C_s}{C} \tag{14.10}$$

where $C_s$ = the solubility of the adsorbate.

If the value of $\epsilon$ can be determined for any solute of a single or multiple component solution, its wet solid phase concentration, $Q_3^0$, can be estimated by the methods described in Chapter XIII. The calculation of $\epsilon$, however, involves the solubility $C_s$, which for organic multicomponent solutions is a complex function of the particular organic concentrations and is often known. Although the experimental data base available is extremely limited, for accurate determinations of water solubility of organic mixtures to be made, group contribution methods like the one proposed by Fredenslund et al. [11] can be used for approximate estimations.

Assuming that the value of $Q_3^0$ has been calculated, the contribution, $Q_3$, of the wet surface to overall capacity is:

$$Q_3 = Q_3^0 A_w \tag{14.11}$$

The definitions and units of $Q_3$ and $Q_3^0$ are analogous to those of $Q_1$ and $Q_1^0$.

Total Organic Adsorbed. The overall adsorption capacity, $Q_T$, for the organic is:

$$Q_T = Q_1 + Q_2 + Q_3 \tag{14.12}$$

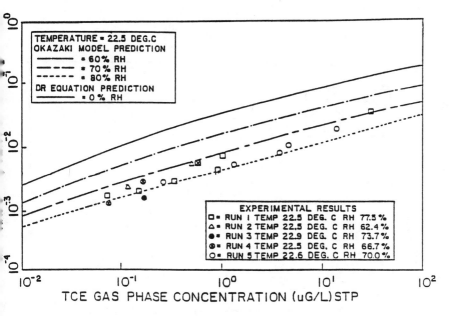

Figure 14.10.    Effect of Relative Humidities on TCE Adsorption
Capacity Based on Water Vapor Isotherm from Freeman and
Reucroft [4]

*Application and Limitations of Okazaki's Model*

This model was used by Tang *et al.* [10] to describe the adsorption
of a single organic vapor (trichloroethylene) at ppb levels from a
humid gas stream. They were able to predict the overall adsorption
capacity at elevated R.H. reasonably well, as shown in Figure 14.10.
They also demonstrated the ability of this model to determine the
optimum temperature at any organic vapor concentration, as shown in
Figure 14.11. They found, however, that the predictions of the model
were very sensitive to the water isotherm used.

The application of   Okazaki's   model has two major limitations.
The first has to do with the currently limited knowledge about the
water solubility of multicomponent organic mixtures.   The lack of
dependable experimental data to determine those solubilities  limits

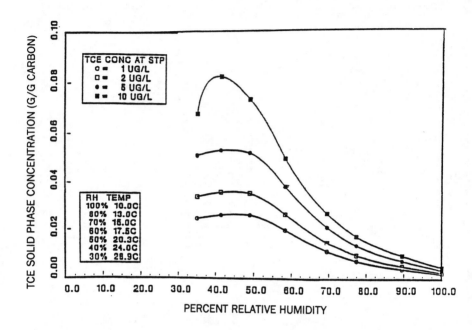

**Figure 14.11.** Effect of Controlling Relative Humidity of Air Stripping Off-Gas on TCE Adsorption Capacity

the accuracy provided by the adsorption potential theory in determining the capacities of the wetted part of the carbon, if more than one organic vapor exists in the stream. Improved characterization of multi-solute aqueous organic solutions through future research, can solve this problem.

The second limitation of this model is an inherent one, stemming from the assumption that the organic vapor does not interfere with the capillary condensation of the water vapor. This assumption is realistic only if the partial pressure of the organic vapor is extremely low, and the humidity approaches saturation. Those conditions are typically met in off-gas stream, resulting from air stripping operations of lightly polluted groundwaters using a high air to liquid ratio.

At higher organic vapor concentrations, however, the extensively adsorbed organic reduces the capacity of the carbon for moisture [7].
As is going to be shown below, VOC concentrations high enough to
interfere with the adsorption of water are common in off-gas streams
resulting from air-stripping of contaminated soils and eventually, of
heavily polluted groundwaters. In those cases Okazaki's model is not
applicable, unless the effect of the organic adsorbates on the adsorption of water is quantified. This is the topic of the next section of this
chapter.

*Water-Organic Adsorbate Interaction*

In an experimental study of vapor adsorption under high humidity
the interference of water with the adsorption of organics and the effect of the organic adsorbates on the adsorption of water vapor were
investigated. The experiments were conducted by passing identical
humid VOC-laden gas streams through a long column and a gravimetric differential reactor. The long column effluent was analyzed
by a flame ionization detector to monitor the VOC adsorption and
determine its adsorption capacity and obtained the total amount of
VOC and moisture adsorbed at equilibrium from the gravimetric measurements. The water uptake was then obtained by subtraction. A
detailed description of each of those two experimental techniques can
be found in Chapters IV (Gravimetric Differential Reactor) and IX
(Long Column).

The compatibility of the differential reactor and the long column
results was examined before combining those two experimental methods. The single vapor isotherm of each of the organics at $25^0$C was
determined by each of the techniques. The results are shown in Table
14.1. The average relative error between the two methods (based on
the gravimetric measurements) was 1.4% and the maximum, 3.7%.

The adsorbent in this experiment was BAC activated carbon
(Union Carbide) and the VOC s tested were trichloroethylene and
toluene. The adsorption equilibria of single component (dry VOC
or organic-free water vapor), as well as of binary systems, with water being one of the adsorbates were studied. The single component
isotherms which were derived are shown in Figure 14.12. The behavior of the binary systems indicated that the organic adsorbate
competes with the water for the same pore volume. As Figures 14.13
and 14.14 show, the capacity for the organic adsorbate is very sensitive to humidity at low VOC concentrations but it gradually becomes
insensitive as the organic vapor concentration is increased. At suffi-

Table 14.1. Comparison of Adsorption Equilibrium Values
Measured by the Quartz Spring and Long Column Methods
on BAC at $25^0$C

| Organics | $C_0$(ppm) | $Q_e$(mg/g) | | Relative Errors |
| | | Quartz Spring | Long Column | |
|---|---|---|---|---|
| Toluene | 170 | 326.9 | 329.2 | 0.7% |
| | 700 | 370.7 | 374.9 | 1.1% |
| | 3550 | 402.1 | 406.1 | 1.0% |
| Trichloro- | 84 | 327.3 | 339.5 | 3.7% |
| ethylene | 834 | 525.7 | 528.2 | 0.5% |
| | 4155 | 608.0 | 600.3 | 1.3% |
| Average Relative Error | | | | 1.4% |

ciently high VOC concentrations, the organic adsorbate dominates the process, occupying most of the adsorption volume and displacing the water.

In those systems, the organic-water interactions could be described by the Lewis correlation [12], which for the case of a binary system is given below:

$$\frac{N_1}{N_1'} + \frac{N_2}{N_2'} = 1 \tag{14.13}$$

where $N_1, N_2$ = mass of organic (1) and moisture (2) adsorbed from the gas mixture, $N_1', N_2'$ = mass of organic (1) and moisture (2) adsorbed in a pure single component system at the same vapor pressure or relative humidity.

This is an empirical relationship first observed to be followed by a multicomponent hydrocarbon mixture [12]. Its binary system form, shown in Equation (14.13), can be expanded to describe equilibria of higher multicomponent systems. For the case of a tertiary system (two organics plus water), the correlation becomes:

$$\frac{N_1}{N_1'} + \frac{N_2}{N_2'} + \frac{N_3}{N_3'} = 1 \tag{14.14}$$

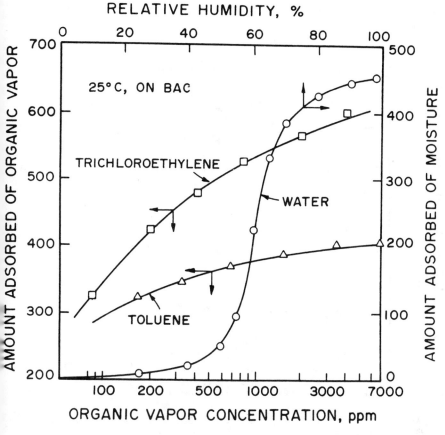

**Figure 14.12.** Isotherm for Toluene Trichloroethylene and Water

where the subscript (3) stands for the second organic. In the same way, additional terms can be added if more adsorbates are present:

$$\sum_{i=1}^{n} \frac{N_i}{N_i^!} = 1 \qquad (14.15)$$

The fit of the water-organic adsorbate experimental data to the Lewis correlation is shown in Figure 14.15. It should be noted that the toluene (hydrocarbon)-water system followed the correlation exactly, while the trichloroethylene (chlorinated hydrocarbon)-water system exhibited a 10% deviation. More data are required to make clear whether this deviation is a random one or a systematic pattern for chlorinated hydrocarbons. In any case, the Lewis correlation seems to be promising and sufficiently accurate for engineering estimations. It can be used along with Okazaki's Model in the following way:

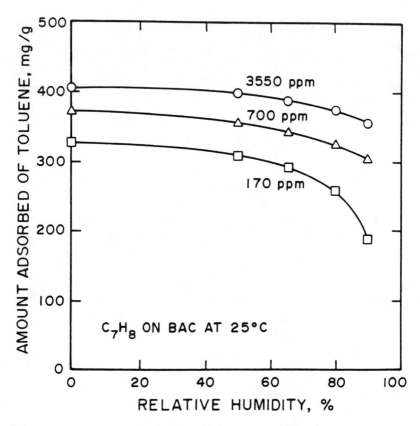

**Figure 14.13.** Amount of Toluene Adsorbed as a Function of Relative Humidity

1. Okazaki's Model is applied in its original form assuming that the amount of water adsorbed $(N_2)$ is as predicted by the water isotherm $(N_1')$.
2. The overall adsorption capacity $(Q_T)$ for each of the organics is calculated as first approximation.
3. The predicted $Q_T$ values are used in the Lewis correlation (as $N_1, N_3$, etc.) to provide a corrected estimate of the amount of water adsorbed $(N_2)$.
4. Okazaki's Model is applied again, using the corrected value of the adsorbed water, and new $Q_T$ values are obtained.
5. Steps 3 and 4 are repeated until two successive estimates of the water adsorbed $(N_2)$ are sufficiently close.

The repetitive approach described above, can correct for the effect of the organic adsorbates on the adsorption of water. Its usage is

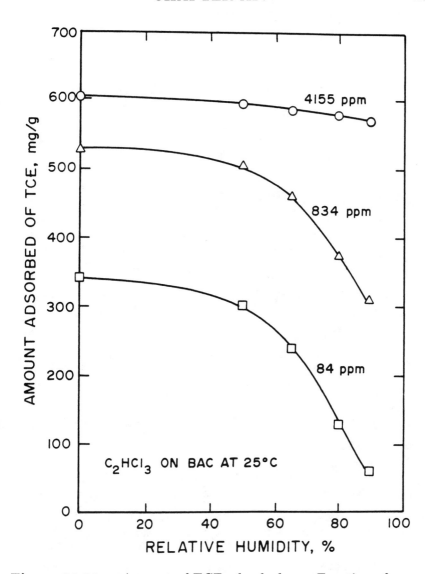

**Figure 14.14.**    Amount of TCE adsorbed as a Function of
  Relative Humidity

recommended in any case where the organic vapors are concentrated
enough to significantly alter the adsorption of water. As the experimental data indicated, a noticeable reduction of the capacity for water
was observed at toluene and trichloroethylene vapor concentrations of
tens of ppm or higher. VOC levels of tens and hundreds of ppm are
typical at the off-gas stream following air stripping of contaminated
soils [3]. Also, concentrations of tens of ppm represent the upper limit

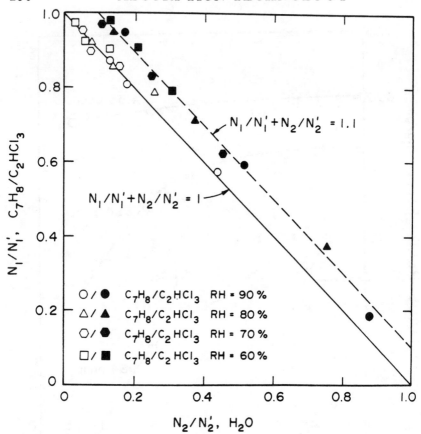

**Figure 14.15.** Correlation between the Quantity of Water and Organic Vapor Adsorbed

for the emissions resulting from air-stripping of groundwaters. VOC s that are adsorbed more readily than those tested by Noll et al. [7] may have a significant effect on the adsorption of water at even lower vapor concentrations. The first cycle of the procedure recommended above permits the engineer to decide whether the reduction of the capacity for water is significant or not for any particular system.

*REFERENCES*

1. Cortright, R.D., J.C. Crittenden, B.G. Rick, S.R. Tang, D.L. Perram, and T.J. Rigg, *Removal of Volatile Organic Chemicals from Air Stripping Tower Off-Gas Using Granular Activated Carbon.* Vol. I. in Crittenden *et al. An Evaluation of the Technical Feasibility of the Air Stripping Solvent Recovery Process.* Research Report, AWWA Research Foundation (1987).

2. Coutant, R.W., T. Zwick, and B.C. Kim, *Removal of Volatile Organics from Humidified Air Streams by Adsorption.* Final Report, Engineering & Services Laboratory, Air Force Engineering & Services Center (1987).

3. Koltuniak, D.L. *Chemical Engineering,* 30 (1986).

4. Brunauer, S. *J. Am. Chem. Soc.,* 62, 1723, (1940).

5. Perry, R. H., Green, D. W., and Maloney, J. O., *Perry's Chemical Engineers' Handbook.* Sixth Edition, McGraw-Hill, Inc., New York, NY (1984).

6. Okazaki, M., H. Tamon, and R. Toei, *J. Chem. Eng. Japan,* 11, 209 (1978).

7. Noll, K.E., D. Wang, and T. Shen, *The Effect of Moisture on the Vapor Phase Adsorption of Toluene and Trichloroethylene.* APCA 81st Annual Meeting, Dallas Texas (1988).

8. Juhola, A.J. and E.O. Wiig, *J. Am. Chem. Soc.,* 71, 2069 (1949).

9. Defay, R., I. Prigonone, A. Bellemans, and D.H. Everett, *Surface Tension and Adsorption.* John Wiley and Sons, Inc., New York 1966.

10. Tang, S.R., J.C. Crittenden, R.D. Cortright, B.G. Rick, D.L. Perram, and T.J. Rigg, *Description of Adsorption Equilibria for Volatile Organic Chemicals in Air Stripping Tower Off-Gas.* Vol. 2. in Crittenden *et al. An Evaluation of the Technical Feasibility of the Air Stripping Solvent Recovery Process.* Research Report, AWWA Research Foundation (1987).

11. Fredenslund, A., R. Jones, and J.M. Prausnitz, *AIChE. J.,* 21 1086 (1975).

12. Lewis W.K., E.R. Gilliland, B. Chertow, and W.P. Cadogan, *Ind. Eng. Chem.,* 42, 1319 (1950).

## A METHOD FOR NONIDEAL ADSORBED PREDICTION FOR LIQUID-SOLID SYSTEMS

Contributing Author: J.M. Wu

### INTRODUCTION

Water is the most popular solvent in use at present. However, in liquid adsorption processes, water is typically treated as a carrier liquid similar to nitrogen in gas adsorption processes. Because of this, few people think of water as an adsorbate but only as a carrier liquid in the design of liquid adsorption systems. It has been shown that water vapor affects the equilibrium adsorption capacity for organics in gas adsorption systems [1]. The higher the humidity and the lower the organic concentration, the lower the adsorption capacity of the organic compound. In liquid adsorption systems, the same effect has not been demonstrated. The purpose of this chapter is to examine the effect of water on the adsorption capacity of organics in liquid adsorption systems. Gravimetric adsorption techniques are employed and a set of equations are developed that allows the calculation of the water adsorption capacity in the presence of organic compounds. This information is then used to calculate an activity coefficient between the water and the organics on the adsorbed phase. Experimental data are reported for adsorption from binary and ternary liquid mixtures of water and other compounds (phenol, $p$-chlorophenol, 1,4-butanediol, 2-methylamino-ethanol) on activated carbon (BAC) at $25^0$C. The equations that are developed allow calculation of the water adsorption capacity for the mixtures. A method is also presented that allows the prediction of the water adsorption capacity for ternary mixtures using data obtained from binary mixtures. This system exhibits nonidealities in the adsorbed phase. Predicted equilibrium data are found to be in quantitative agreement with the experimental values.

### THEORY DEVELOPMENT

*Adsorbed Phase and Liquid Phase Surface Excess*

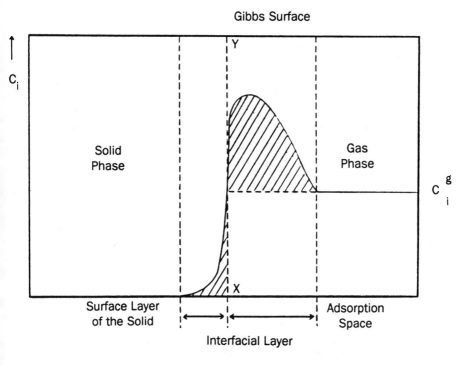

**Figure 15.1.**    Adsorption at the Solid-Gas Interface

The surface excess amount of the adsorbed gas, $n_i^e$, is initially defined by the excess number of that substance present in the system over the number present in a reference system where adsorption does not occur at the same equilibrium gas pressure. As shown in Figure 15.1 for the adsorption at the solid-gas interface, the surface excess can be expressed as [2]:

$$n_i^e = (C_i' - C_i^g)dV + C_i'dV \tag{15.1}$$

where $C_i'$ = the local concentration of substance $i$ in a volume element $dV$ of the interfacial layer, $C_i^g$ = the concentration of that substance in the bulk of the gas phase.

The first term of Equation (15.1) is adsorption space, and the second term, surface layer of adsorption, is usually assumed to be zero; therefore,

$$n_i^e = (C_i' - C_i^g)dV \tag{15.2}$$

For a liquid-solid adsorption system, the surface excess can be obtained by measurements of the isothermal change in composition of the bulk liquid ($\Delta X$) when the solid adsorbent is immersed in the liquid mixture:

$$n_i^e = n^0(X_i^0 - X_i) \qquad (15.3)$$

where $n^0$ = the total mole in the original liquid solution, $X_i^0, X_i$ = the mole fractions of the $i$th component before and after the adsorbent is immersed in the liquid, respectively.

We can assume that the total moles, $n^0$, in the original liquid solution is unchanged and can be divided into two phases: an adsorbed phase containing $n'$ moles and a bulk phase containing $n$ moles, so that

$$n^0 = n + n' \qquad (15.4)$$

Equation (15.4) is a material balance on the total number of moles of liquid. Combining Equations (15.3), (15.4), and a material balance on the $i$th component of the mixture $n^0 X^0 = nX_i + n'X_i'$, the surface excess in terms of properties of the adsorbed phase can be expressed as:

$$n_i^e = n'(X_i' - X_i) \qquad (15.5)$$

Equation (15.5) shows that the surface excess is the amount of $i$ adsorbed less than would be adsorbed, if the composition of the adsorbed phase were the same as that of the liquid phase.

The value of $\Delta X$ given by Equation (15.5) is the ideal value corresponding to complete removal of the bulk liquid from the adsorbent. It follows from Equation (15.3) that

$$\sum n_i^e = 0 \qquad (15.6)$$

In the special case of strongly adsorbed solutions for which $X_i' \gg X_i$, $n_i^e \approx n'X'$ by Equation (15.5). For this special case, the surface excess is simply the amount adsorbed.

For microporous adsorbents such as activated carbon, the total moles on the adsorbed phases, $n'$, may be estimated by

$$\frac{1}{n'} = \sum \frac{X_i'}{m_i} \qquad (15.7)$$

where $m_i$ = the capacity of pores in moles for pure $i$th liquid given by the amount adsorbed from the vapor of that liquid at saturation. Equation (15.7) is based on the reasonable assumption that the volume change is zero when the solution is formed in the micropores from the pure liquid. $m_i$ can be obtained either by experiment, if the adsorbate is volatile, or, according to Gurvitsch's rule [3], where $m_i$ is approximately proportional to the bulk liquid density of the adsorbate, that is

$$m_i = \frac{V_p}{v_i} \qquad (15.8)$$

where $V_p$ = the micropore volume of the adsorbent, and $v_i$ = the molar volume of adsorbate $i$.

By using the gravimetric technique as the experimental method, information can be obtained on the mass of organic adsorbate adsorbed, $W$. Therefore, the total amount adsorbed on the adsorbed phase is written as:

$$n' = \frac{W_i}{X'_i} \qquad (15.9)$$

The definition of mole fraction on the adsorbed phase results in Equation (15.10):

$$\sum_i X'_i = 1 \qquad (15.10)$$

Equations (15.5), (15.7), (15.9), and (15.10) are all derived based on the adsorbed phases. By solving these equations for a binary system, we can obtain the information required to fully characterize the system, such as the surface excess, $n_i^e$, the total moles on adsorbed phase, $n'$, and the adsorbed phase mole fraction, $X'_i$. Those results may be checked by comparing the surface excess with Equation (15.3), which is the surface excess obtained from the experiments. Theoretically, the surface excess obtained from the adsorbed phase and the liquid phase (the experiment) should be equal.

*Prediction Model for Ternary Liquid Solutions*

A predictive model for multicomponent liquid mixtures (ternary and higher) in terms of adsorption from pairs of liquids was first derived by Minka and Myers [4]. The most important application of this model is the special case of two dilute solutes in a solvent. If the solvent is water and two organics are solutes, it is possible to predict the adsorption capacity for these compounds.

At equilibrium, the fugacity in the bulk phase $f_i$ is equal to the fugacity in the adsorbed phase $f'_i$:

$$f_i = f'_i \qquad (15.11)$$

The fugacity in the bulk phase [5] is:

$$f_i = P_i^s \gamma_i X_i \exp\left[\frac{v_i^s(P - P_i^s)}{m_i RT}\right] \qquad (15.12)$$

Similary, the fugacity in the adsorbed phase [6] is

$$f_i' = P_i^s \gamma_i' X_i' \exp\left[\frac{-(\phi - \phi_i^0)}{m_i RT}\right] \qquad (15.13)$$

where $\phi_i^0$ = the free energy of immersion of the adsorbent in the pure liquid phase, $\phi$ = the free energy of immersion in the mixture, $\gamma_i = yP_i/X_iP_i^s$ the activity coefficient. For most adsorption systems the pressure is subatmosphere and the exponential term in Equation (15.12) may be equal to unity with neligible error. Equations (15.11) to (15.13) then give

$$\gamma_i X_i = \gamma_i' X_i' \exp\left[\frac{-(\phi - \phi_i^0)}{m_i RT}\right] \qquad (15.14)$$

The variables $n'$ and $X_i'$ are eliminated from Equation (15.5) using Equations (15.7) and (15.14). The result, after some algebraic manipulation, is

$$n_i^e = \frac{\sum X_i X_j (1 - K_{ij})}{\sum X_i K_{ij}/m_j} \qquad (15.15)$$

and

$$K_{ij} = \frac{\gamma_i' \gamma_j}{\gamma_j' \gamma_i} \exp\left[\frac{\phi - \phi_j^0}{m_j RT} - \frac{\phi - \phi_i^0}{m_i RT}\right] \qquad (15.16)$$

where $K_{ij}$ = a function of composition. The activity coefficients in the adsorbed phase $(\gamma_i', \gamma_j')$ are a function of the composition of the adsorbed phase $(X_i', X_j')$. This provides a general correlation allowing estimation of the $X_j'$ from the bulk liquid phase composition [4]:

$$X_j' = \frac{X_j K_{ij}}{\sum X_j K_{ij}} \qquad (15.17)$$

The summation in Equations (15.15) and (15.17) are over $j = 1, 2, 3, \cdots, n$. The quantity $K_{ij}$ has the following properties:

$$K_{ij} = 1 \quad \text{as } i = j; \quad K_{ij} = \frac{1}{K_{ji}}; \quad K_{ij} = K_{ip}K_{pj} \quad (15.18)$$

The free energy of immersion ($\phi$) is also a function of composition and is obtained by integrating the Gibbs adsorption isotherm [6]:

$$-d\phi = n' d\mu_i \tag{15.19}$$

For a ternary liquid solution, Equation (15.6) can be extended as:

$$n_1^e + n_2^e + n_3^e = 0 \tag{15.20}$$

Two of the surface excesses can be determined by Equation (15.15):

$$n_1^e = \frac{X_1 X_2 (1 - K_{12}) + X_1 X_3 (1 - K_{13})}{X_1/m_1 + X_2 K_{12}/m_2 + X_1 K_{13}/m_3} \tag{15.21}$$

$$n_2^e = \frac{X_1 X_2 (1 - K_{21}) + X_2 X_3 (1 - K_{23})}{X_1 K_{21}/m_1 + X_2/m_2 + X_3 K_{23}/m_3} \tag{15.22}$$

and

$$K_{12} = \frac{\gamma_1' \gamma_2}{\gamma_2' \gamma_1} \exp\left[\frac{\phi - \phi_2^0}{m_2 RT} - \frac{\phi - \phi_1^0}{m_1 RT}\right] \tag{15.23}$$

$$K_{13} = \frac{\gamma_1' \gamma_3}{\gamma_3' \gamma_1} \exp\left[\frac{\phi - \phi_3^0}{m_3 RT} - \frac{\phi - \phi_1^0}{m_1 RT}\right] \tag{15.24}$$

$$K_{21} = \frac{1}{K_{12}}; \qquad K_{23} = \frac{K_{13}}{K_{12}} \tag{15.25}$$

The chemical potentials in the bulk liquid are related by the isothermal Gibbs-Duhem equation [5]:

$$X_1 d\mu_1 + X_2 d\mu_2 + X_3 d\mu_3 = 0 \tag{15.26}$$

From Equation (15.5),

$$n' = n_i^e + n' X' \tag{15.27}$$

Substitution of Equations (15.20) to (15.27) into (15.19) followed by algebraic rearrangement gives:

$$d\phi = \frac{X_2 (1 - K_{12}) d\mu_2 + X_3 (1 - K_{13}) d\mu_3}{X_1/m_1 + X_2 K_{12}/m_2 + X_3 K_{13}/m_3} \tag{15.28}$$

Integrating along the path, $X_3/X_2$ = constant, gives

$$\frac{\phi - \phi_1^0}{RT} = \int_{X_1=1}^{X_1} f(X_1) dX_1 \tag{15.29a}$$

and

$$f(X_1) =$$

$$\frac{X_2(1 - K_{12})\left(\dfrac{d\ln\gamma_2}{dX_1} - \dfrac{1}{1-X_1}\right) + X_3(1 - K_{13})\left(\dfrac{d\ln\gamma_3}{dX_1} - \dfrac{1}{1-X_1}\right)}{X_1/m_1 + X_2 K_{12}/m_2 + X_3 K_{13}/m_3}$$

$$(15.29b)$$

According to Equation (15.17)

$$X_1' = \frac{X_1}{X_1 + X_2 K_{12} + X_3 K_{13}}; \quad X_2' = \frac{X_2 K_{12}}{X_1 + X_2 K_{12} + X_3 K_{13}} \quad (15.30)$$

After numerical integration of Equation (15.29) for $\phi$, the surface excess is given by Equations (15.20) to (15.25).

A flow diagram for prediction of multicomponent adsorption equilibrium capacity from binary mixtures is shown in Figure 15.2.

## EXPERIMENTAL METHOD

A differential reactor adsorption system employing the recycle mode was set up as shown in Figure 15.3. The total volume of the liquid in the system was about 1500 ml. Four organic compounds (phenol, $p$-chlorophenol, 1,4-butanediol, 2-methylamino-ethanol) were selected as the adsorbates. Activated carbon (BAC) made by Union Carbide Corporation was used as the adsorbent. The characteristics and physical properties of these adsorbates are shown in Table 15.1. The room temperature was kept at $25 \pm 1^0 C$. After recycling the solutions, equilibrium was determined by measuring the extension of the Quartz Spring until it no longer changed. Concentrations of the solution were measured by a UV Spectrophotometer for comparison with the results obtained from the gravimetric technique.

After equilibrium was reached, the organic adsorbate adsorbed amount $W_i$ was obtained by recording the total extension of the Quartz Spring and correcting for buoyance [7]. The equilibrium concentrations for the organic compounds were measured by UV Spectrophotometer. With $W_i$ determined from the experiments described above, Equations (15.5), (15.7), (15.9), and (15.10) become four equations with four unknowns $(n_i^e, n', X_1', X_2')$ that can be solved mathematically. The surface excess, $n_i^e$, obtained from these calculations can then be compared with those obtained from Equation (15.1) from the experiments. Calculated parameters are then used to predict the adsorption equilibrium capacity for a ternary system.

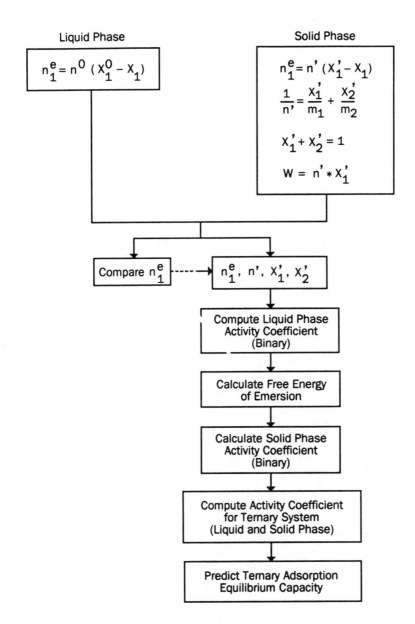

**Figure 15.2.**    Diagram in Prediction of Multicomponent Adsorption Equilibrium Capacity

*RESULTS AND CONCLUSIONS*

Table 15.1. Characteristics of the Adsorbates

| Properties | Phenol | p-Chloro-phenol | Water | 1,4-Butane-diol | 2-Methyl-amino-ethanol |
|---|---|---|---|---|---|
| M.W.(g/mole) | 94 | 128 | 18 | 90.12 | 75 |
| $P_c$(atm)[11] | 695.1 | 756.9 | 581 | 477.2 | 612.4 |
| Solubility[11] Parameter | 9.92 | 9.06 | 18.44 | 7.73 | 8.28 |
| Buoyancy[7] Factor | 2.06 | 1.81 | – | 2.42 | 5.61 |
| Max. Capacity[3] (mmole/g) | 5.36 | 4.64 | 23.9 | 5.08 | 5.61 |
| Density (liq.) (g/cm³) | – | – | 1.0 | 1.017 | 0.935 |
| Molar Volume (cm³/mole) | 87.8 | 101.6 | 18 | 88.5 | 80.2 |

*Calculation of Activity Coefficient in the Bulk Phase*

The following equation proposed by Hildebrand [8] for high pressure vapor-liquid equilibria and successfully applied for liquid-liquid solutions of hydrocarbons [9] was used in calculation of the bulk phase activity coefficient:

$$\ln \gamma_i = \frac{v_i(\delta - \bar{\delta})^2}{RT} \qquad (15.31)$$

This equation gives the activity coefficient as a function of temperature and composition. It requires two constants for each component: the solubility parameter $\delta$ and liquid molar volume $v_i$. The quantity $\bar{\delta}$ designates as an average value of the solubility parameter for the solution:

$$\bar{\delta} = \frac{\sum X_i v_i \delta_i}{\sum X_i V_i} \qquad (15.32)$$

The solubility parameter $\delta_i$ and liquid molar volume $v_i$ for different compounds are given in Table 15.1. Therefore, the activity coefficient in the bulk phase as a function of the liquid phase mole fraction can be calculated.

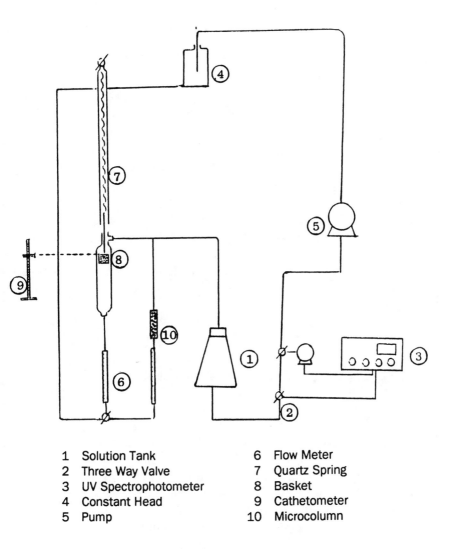

| 1 | Solution Tank | 6 | Flow Meter |
|---|---|---|---|
| 2 | Three Way Valve | 7 | Quartz Spring |
| 3 | UV Spectrophotometer | 8 | Basket |
| 4 | Constant Head | 9 | Cathetometer |
| 5 | Pump | 10 | Microcolumn |

**Figure 15.3.** Experimental Apparatus for Adsorption Equilibrium

*Calculation of Activity Coefficient in the Adsorbed Phase*

The activity coefficients for the binary solutions in the adsorbed phase were calculated from the surface excess obtained from the experiments. For component 1, the combination of Equations (15.17) and (15.23) with $X_3 = 0$ gives the composition of the adsorbed solution:

$$X_1' = \frac{m_1(m_2 X_1 + n_1^e)}{m_1 m_2 + n_1^e(m_1 - m_2)} \tag{15.33}$$

The free energy of immersion, from Equation (15.29) with $X_3 = 0$ becomes:

$$\frac{\phi - \phi_1^0}{RT} = -\int_{X_1=1}^{X_1} \frac{n_1^e}{X_1 X_2 \gamma_1} d(\gamma_1 X_1) \tag{15.34}$$

Then, from Equation (15.4),

$$\gamma_1' = \frac{X_1}{X_1'}\gamma_1 \exp\left(\frac{\phi - \phi_1^0}{m_1 RT}\right), \quad \gamma_2' = \frac{X_2}{X_2'}\gamma_2 \exp\left(\frac{\phi - \phi_2^0}{m_2 RT}\right) \tag{15.35}$$

Given the experimental isotherms and the surface excess as a fraction of the liquid phase mole fraction, the calculation of the activity coefficient in the adsorbed phase is straightforward. After numerical integration of Equation (15.35), the activity coefficients in the adsorbed phase as a function of composition, $\gamma_1'(X_1'), \gamma_2'(X_2')$, were calculated from Equations (15.33) and (15.35).

Several models such as Margues, NRTC, Wilson [5] and UNIQUAC [10] have been used to describe the activity coefficient. For moderately nonideal binary mixtures, all of the equations that contain two (or more) binary parameters give good results, and there is little reason to choose one over another. But for strongly nonideal binary mixtures, the Wilson equation is probably the most useful, because, unlike the NRTC equation, it contains only two adjustable parameters and it is mathematically simpler than the UNIQUAC equation [11]. Therefore, the Wilson equation will be used to describe the activity coefficients for the binary solutions. The best values for the two parameters in the Wilson equation can be found by numerical regression techniques based on a least-mean-square fit of the data. These two parameters can then be used to calculate the activity coefficients for the ternary system.

The surface excess values obtained from the experiments and from the calculations are compared in Figures 15.4 to 15.8, respectively. The figures show that for the experimental concentration range used here, the surface excess obtained by the two methods are similar. This means that Equations (15.5), (15.7), (15.9), and (15.10) may be applied to calculate the surface excess and also the other unknown parameters in the equations. The surface excess shown in Figures 15.6 to 15.8 for the water, 1,4-butanediol and 1-methylamino-ethanol system reveals that 1,4-butanediol is the strongest adsorbate, since there

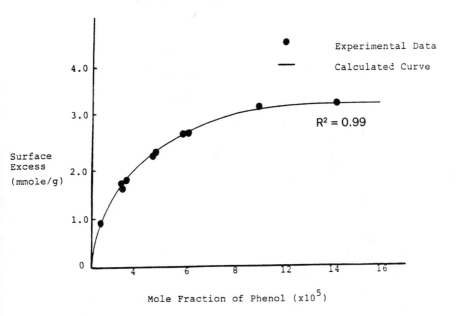

**Figure 15.4.**    Surface Excess of Phenol in Water

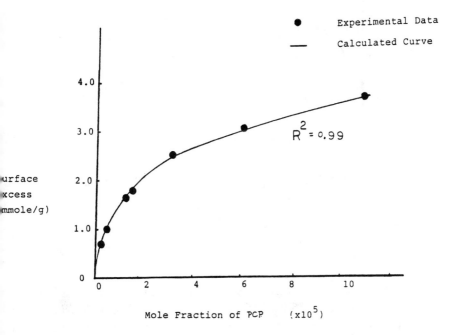

**Figure 15.5.**    Surface Excess of PCP in Water

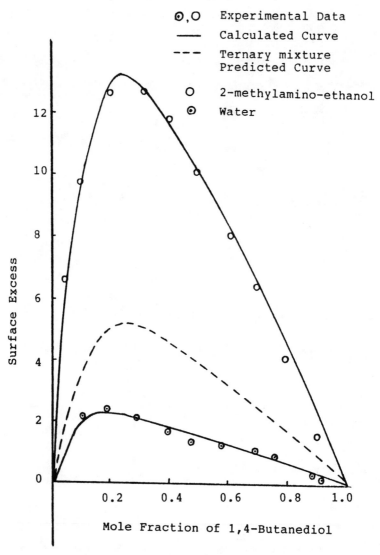

**Figure 15.6.**    Surface Excess of 1,4-Butanediol

are positive surface excesses with both of the other two adsorbates. Water is the second strongest adsorbent and 2-methylamino-ethanol is the weakest.

The adsorbed phase mole fractions for all of the adsorbates used in this study were obtained by solving the four equations and the results are shown in Figures 15.9 to 15.11, respectively. The mole fraction of phenol increased from zero to 0.25 as the equilibrium concentration was increased from zero to about 10 mmole/l. At the highest concen-

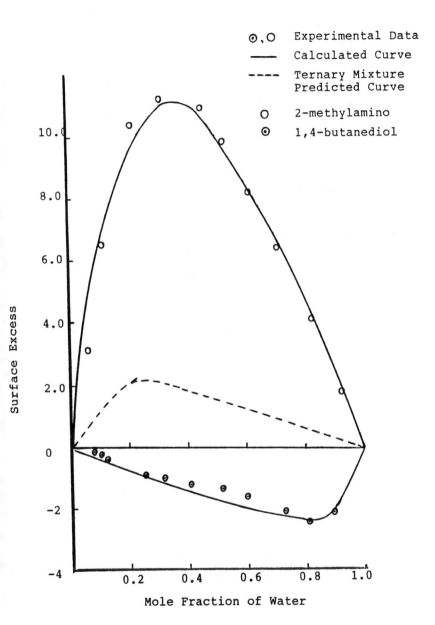

Figure 15.7.    Surface Excess of Water

tration (10 mmole/l), only about 0.25 mole fraction was phenol and the other available adsorbing area was occupied by water. The adsorbed water on the adsorbed phase was three times greater than the adsorbed phenol. Figure 15.10 shows similar results. The mole fraction

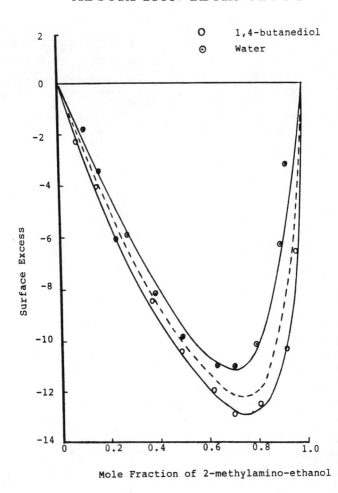

**Figure 15.8.**     Surface Excess of 2-Methylamino-Ethanol

of PCP rapidly increased from zero to about 0.5 as the equilibrium concentration increased from zero to about 4.25 mmole/l. At this concentration (4.25 mmole/l), the mole fraction of water adsorbed was the same as that of PCP.

Because of their limited solubility, phenol and PCP were replaced by 1,4-butanediol and 2-methylamino-ethanol, which are completely miscible with water, to provide experimental data for higher concentrations. The solid phase mole fraction of water, Figure 15.11, was higher than for 2-methylamino-ethanol until the liquid phase mole fraction of organic exceeded 0.7. Since 1,4-butanediol was a stronger sorbate than 2-methylamino-ethanol, the solid phase mole fraction for water can be seen to decrease faster in the presence of 1,4-butanediol.

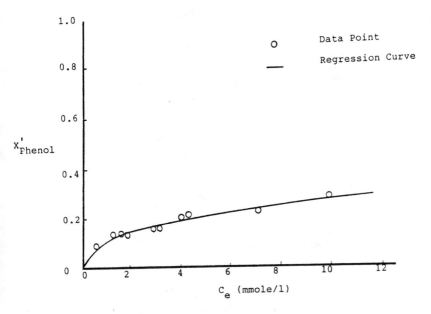

**Figure 15.9.**   Solid Phase Mole Fraction vs. Equilibrium
Concentration for Phenol/BAC

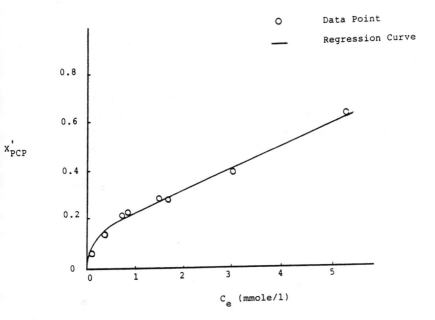

**Figure 15.10.**   Solid Phase Mole Fraction vs. Equilibrium
Concentration for *p*-Chlorophenol/BAC

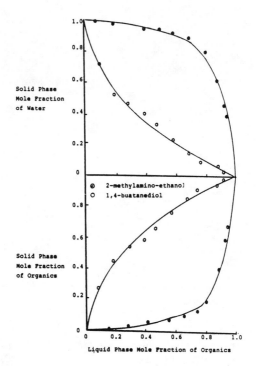

**Figure 15.11.** Mole Fractions in the Solid Phase and Liquid Phase

Even so, the amount of water adsorbed was still relatively high and decreased as the liquid phase mole fraction of 1,4-butanediol increased.

It would appear from the data in Figures 15.9 to 15.11 that water occupied a large portion of the available adsorbing area for these four adsorbates tested in this study. To quantify the amount of adsorbed water, equilibrium activity coefficients for both of the binary solutions both in the adsorbed phase and in the bulk phase as a function of composition, $\gamma_1(X_1), \gamma_2(X_2), \gamma_1'(X_1'), \gamma_2'(X_2')$, were calculated from Equations (15.31) and (15.35). The results are plotted in Figures 15.12 to 15.16. The activity coefficients for the adsorbed phase show a complex variation with equilibrium composition and there is no obvious relationship to the nonideal behavior observed for the bulk liquid solutions. The ratio of activity coefficients between the organic compounds and water in the adsorbed phase changes as a function of equilibrium composition and is quite different than the activity coefficients for the same equilibrium composition in the bulk phase. This result demonstrates that the water adsorbed on the solid phase affects the equilibrium capacity of the organic compounds.

Activity coefficient constants from the Wilson equation for each

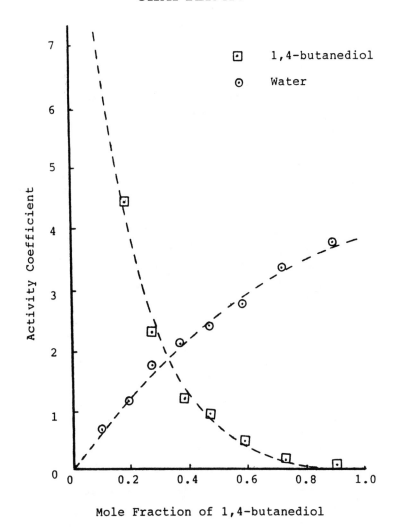

**Figure 15.12.**   Activity Coefficients of Water and 1,4-Butanediol in the Solid Phase

compound are reported in Table 15.2. For different compounds at the same composition, the ratios of the activity coefficient are different. It appears that the effect of adsorbed water is different for different organic compounds. Therefore, it may be necessary to consider water as a separate compound for each mixture rather than treat it as a uniform carrier liquid as is done with nitrogen in gas adsorption studies.

The mathematical model that has been described in Equations (15.20) to (15.29) was applied to predict the multicomponent equi-

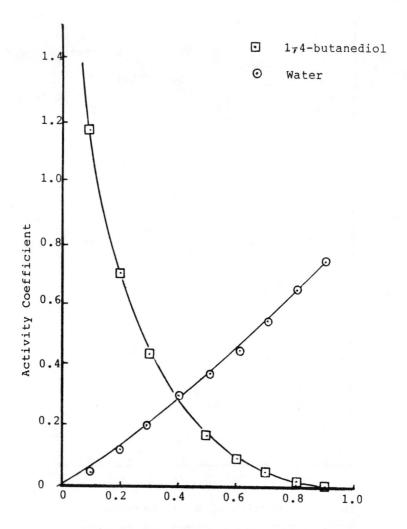

□   1,4-butanediol

⊙   Water

Mole Fraction of 1,4-butanediol

**Figure 15.13.**   Activity Coefficients of Water and 1,4-Butanediol
in the Solid Phase

librium capacity. By arbitrarily taking different compositions of the
three adsorbates, the surface excess obtained from equilibrium experiments and from the calculations are compared in Table 15.3. The
experimental points conform to the predictions within an acceptable
range of 4.1% average error.

This agreement allows us to study the behavior of ternary systems.
The dashed lines in Figures 15.6 to 15.8 are the surface excess for

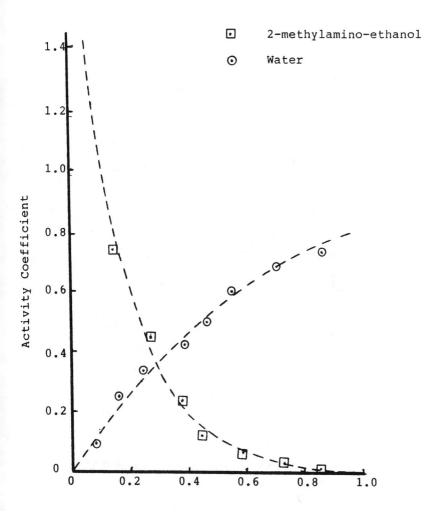

Figure 15.14.   Activity Coefficients of Water and 2-Methylamino-
Ethanol in the Liquid Phase

one of the adsorbates (adsorbate A) adsorbed from a ternary solution
(solution A+B+C) based on equimolar amounts for the other two adsorbates $(X_B = X_C)$. At a fixed mole fraction for any component, the
variation of surface excess with composition is highly   nonlinear. As
would be expected, the dashed lines lie between the surface excess for
the two binary   (solid lines) systems. Thus, for a given mole fraction
of adsorbate in the bulk phase, the surface excess of this adsorbate

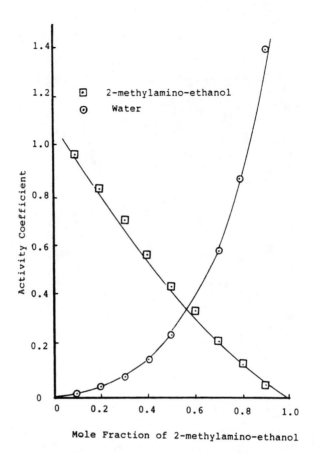

**Figure 15.15.**    Activity Coefficients of Water and 2-Methylamino-
Ethanol in the Solid Phase

goes through a maximum as the molar ratio of the other two adsorbates (weaker adsorbate/stronger adsorbate) increases. For example,
the surface excess for water goes through a maximum as the molar
ratio of 2-methylamino-ethanol to 1,4-butanediol increases.

## CONCLUSIONS

Some conclusions can be summarized from the above analysis as
follows:
1. Calculation of adsorbed phase parameters and water adsorption capacity: The agreement between the surface excess obtained experimentally from the liquid phase and calculated theoretically from
   the solid phase demonstrates that the four equations derived for the

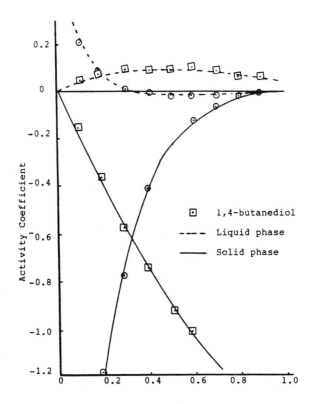

**Figure 15.16.**    Activity Coefficients of 1,4-Butanediol and
2-Methylamino-Ethanol in the Solid and Liquid Phases

solid phase can be applied to the prediction of surface excess. However, not only the surface excess, but also the total moles adsorbed and solid phase mole fraction for each adsorbate can be calculated from those equations. This theory also provides a way to calculate the water adsorption capacity at equilibrium for organic adsorption systems which has not been available heretofore.

2. Effect of water adsorption capacity: The results of this study prove the initial hypothesis of this study, i.e. that water is adsorbed and occupies a large amount of the available solid phase in liquid adsorption systems, especially when the organic liquid phase equilibrium concentration is relatively low. The solid phase activity coefficient calculated from the set of thermodynamic consis-

**Table 15.2.** Constants of Wilson Equation for Binary Mixtures

| Phase | $G_{ij}$ | 1,4-Butanediol versus Water | Water versus 2-Methylamino-ethanol | 2-Methylamino-ethanol versus 1,4-Butanediol |
|---|---|---|---|---|
| Bulk | $G_{12}$ | 0.97 | 0.037 | 3.1 |
| | $G_{21}$ | 0.28 | 2.16 | 0.001 |
| Adsorbed | $G_{12}$ | 0.002 | 1.04 | 228.8 |
| | $G_{21}$ | 0.47 | 0.39 | 0.08 |

**Table 15.3.** Ternary Adsorption Data for the Mixture of 1,4-Butanediol(1)+2-Methylamino-ethanol(2)+Water(3)

| Molar Ratio | | | Excess (mmole/g) | | | | | |
|---|---|---|---|---|---|---|---|---|
| $X_1$ | $X_2$ | $X_3$ | $n_1^e$(exp) | $n_1^e$(calc) | $n_2^e$(exp) | $n_2^e$(calc) | $n_3^e$(exp) | $n_3^e$(calc) |
| 0.09 | 0.84 | 0.07 | 6.15 | 5.98 | -1.82 | -1.87 | -3.96 | -4.11 |
| 0.15 | 0.52 | 0.33 | 5.87 | 5.75 | -10.47 | -10.66 | 4.88 | 4.91 |
| 0.25 | 0.4 | 0.35 | 5.5 | 5.56 | -8.32 | -8.30 | 2.59 | 2.74 |
| 0.33 | 0.31 | 0.36 | 5.2 | 5.0 | -7.23 | -7.09 | 2.11 | 2.09 |
| 0.47 | 0.41 | 0.22 | 6.8 | 6.65 | -8.40 | -8.56 | 1.82 | 1.91 |
| 0.52 | 0.1 | 0.38 | 2.91 | 2.94 | -2.19 | -2.07 | -0.15 | -0.13 |
| 0.69 | 0.15 | 0.16 | 2.31 | 2.29 | -3.62 | -3.45 | 1.07 | 1.16 |
| 0.71 | 0.05 | 0.24 | 4.87 | 4.79 | -1.19 | -1.21 | -3.42 | -3.58 |
| 0.82 | 0.1 | 0.08 | 1.60 | 1.77 | -2.12 | -2.33 | 0.55 | 0.56 |
| 0.88 | 0.07 | 0.05 | 0.98 | 1.19 | -1.77 | -1.52 | 0.37 | 0.33 |

tent equations reveals that adsorbed water has a large effect on the adsorption equilibrium capacity for the organic compounds. The extent of this effect was not the same for 1,4-butanediol and 2-methylamino-ethanol. Hence, taking the water as a "blank" to predict the multicomponent adsorption equilibrium capacity as is usual in the IAS theory may not be adequate for these ranges of concentration.

3. Calculation of activity coefficient for ternary mixture: The agreement between the experimental activity coefficient data and the Wilson equation for both the bulk and the adsorbed phase indicates that the Wilson equation can be applied to describe the activity coefficient in both phases, and hence, can be applied to calculate the activity coefficient for a ternary mixture.

4. Prediction of multicomponent adsorption equilibrium capacity: Quantitative agreement was obtained between model predicted and experimental observed data for a ternary liquid mixture of water, 1,4-butanediol, and 2-methylamino-ethanol. This system is nonideal in both the bulk and the adsorbed phases. This result suggests that multicomponent adsorption may be predicted from binary adsorption.

The significance of this work lies not only in reducing the study of multicomponent adsorption to the study of adsorption for binary solutions, but also in the generality of the theory. The present method is applicable to nonideal adsorption from nonideal solutions at all compositions. In many cases, the IAS model may not be applicable because the solvent cannot be taken as a "blank". In these cases, considering water as one of the solvents allows one to predict the multicomponent adsorption equilibrium capacity.

## REFERENCES

1. Noll, K.E., D.H. Wang, and T. Shen, *The Effect of Relative Humidity on the Vapor Phase Adsorption of $C_7H_8$ and $C_2HCl_3$*. Research Report to IWERC (1987).
2. Manual of Difinitions, *Pure and Appl. Chem.*, **31**, 577 (1972).
3. Gurvitsch, L. *J. Phys. Chem. Soc. Russ.*, **47**, 805 (1915).
4. Minka, C., and A.L. Myers, *AIChE. J.*, **19**, 453 (1973).
5. Prausnitz, J.M. *Molecular Thermodynamics of Fluid-Phase Equilibria*. Prentice-Hall, Englewood Cliffs, New Jersey (1969).
6. Larionov, O.D., and A.L. Myers, *Chem. Eng. Sci.*, **26**, 1025 (1971).
7. Gounaris, V. *Gravimetric Solid Phase Analysis of Adsorption of Phenolic Compounds*. M.S. Thesis, Illinois Institute of Technology (1986).
8. Hildebrand, J.H. and R.L. Scoot, *The Solubility of Non-electrolytes*. 3rd ed., Reinhold, New York (1950).
9. Chao, K.C., and J.D. Seader, *AIChE. J.*, **7**, 598 (1961).
10. Abrams, D.S., and J.M. Prausnitz, *AIChE. J.*, **21**, 116 (1975).
11. Reid, R.C., J.M. Prausnitz, and T.K. Sherwood, *The Properties of Gases and Liquids*. 3rd ed., McGraw-Hill Inc. (1977).

# APPENDIX

---

# MODEL PROGRAM

---

Contributing Author: C. Arai

```
C      ************** FILE NAME : COLM22.FTN **************
C
C      This program gives the breakthrough curve for multi-
C      component systems.  For the detail, a "Design of
C      Multi-Component Long Column Program COLM2.FTN Series"
C      is available per request.  September 20, 1986
C      by C. Arai in IIT, Chicago.
C
C         **********     FILE NAME : COLM22.FTN     **********
C
       IMPLICIT REAL*8(A-H,O-Z)
       INTEGER JUDGE,MODELD,NCOMP,MC,NC,ND,ICOUNT,NND,NE,N1
       INTEGER ITITLE,ISON,MODELI,NDEG,NRESLT,NSTEP,INDEX,I,J,K,M,N
       REAL*8 CB0,CI,WM,AU,B,W,ST,EDS,EDP,DOX,DG,XK,XN,COT,QOT,DELTA
       REAL*8 Y0,SKF,DS,DP,COUT,RP,RHOP,DIA,BL,WT,FLRT,DSTEP,DTOTL,H0,EPS
      +,AREA,BEDVOL,EBED,US,TAU,EON,TSTEP,TOUT,TTOTL,T0,T,OI,TOTL,TOI,TCI
       DIMENSION Y0(126),SKF(3),DS(3),DP(3),COUT(3),OI(3),ITITLE(20),ISON
      +(3)
       COMMON /RTRN/  JUDGE,MODELD,NCOMP
       COMMON /SISO/  CB0(3),CI(3),WM(3)
       COMMON /DIFF/  AU(8,8),B(7,7),W(7),ST(3),EDS(3),EDP(3),DOX(3),DG,MC
      +,NC,ND,ICOUNT
       COMMON /SEQU/  XK(3),XN(3),COT,QOT,DELTA,NND,NE,N1
C
       JUDGE=0
       READ(6,100)(ITITLE(I),I=1,20),MODELI,MODELD,NCOMP,NDEG,NRESLT
       READ(6,110)RP,RHOP,DIA,BL,WT,FLRT,DSTEP,DTOTL,H0,EPS
       READ(6,120)(SKF(I),I=1,NCOMP)
       READ(6,120)(CB0(I),I=1,NCOMP)
       READ(6,120)(WM(I),I=1,NCOMP)
       READ(6,120)(DS(I),I=1,NCOMP)
       READ(6,120)(DP(I),I=1,NCOMP)
       WRITE(7,500)(ITITLE(I),I=1,20)
       WRITE(7,510)MODELI,MODELD,NCOMP,NDEG
       WRITE(7,520)RP,RHOP,DIA,BL,WT,FLRT,DSTEP,DTOTL,H0,EPS
       WRITE(7,530)(SKF(I),I=1,NCOMP)
       WRITE(7,540)(CB0(I),I=1,NCOMP)
       WRITE(7,550)(WM(I),I=1,NCOMP)
       WRITE(7,560)(DS(I),I=1,NCOMP)
       WRITE(7,570)(DP(I),I=1,NCOMP)
       IF(MODELI.NE.1)GO TO 40
       READ(6,130)(ISON(I),I=1,NCOMP)
       M=1
       DO 10 I=1,NCOMP
        IF(ISON(I).GT.M)M=ISON(I)
    10 CONTINUE
```

```
        IF(M.NE.1)GO TO 42
        READ(6,120) (XK(I),I=1,NCOMP)
        READ(6,120) (XN(I),I=1,NCOMP)
        WRITE(7,590) (XK(I),I=1,NCOMP)
        WRITE(7,600) (XN(I),I=1,NCOMP)
        DO 11 K=2,NDEG
         IF(K.EQ.2)READ(5,140)N
         IF(K.NE.2)READ(5,150)N
         IF(K.NE.N)GO TO 41
         MC=K+2
         NC=K+1
         DO 12 I=1,MC
    12    READ(5,160) (AU(I,J),J=1,MC)
         READ(5,170) (W(I),I=1,NC)
         DO 13 I=1,NC
    13    READ(5,160) (B(I,J),J=1,NC)
    11 CONTINUE
C       WRITE(7,650)
C       DO 14 I=1,MC
C   14  WRITE(7,160) (AU(I,J),J=1,MC)
C       WRITE(7,660)
C       WRITE(7,170) (W(I),I=1,NC)
C       WRITE(7,670)
C       DO 15 I=1,NC
C   15  WRITE(7,160) (B(I,J),J=1,NC)
C
        AREA=3.14159D+00*(DIA**2)/4.0D+00
        BEDVOL=BL*AREA
        EBED=(BEDVOL-WT/RHOP)/BEDVOL
        US=FLRT/(AREA*6.0D+01)
        TAU=EBED*BL/US
        TOTL=0.0D+00
        DO 20 I=1,NCOMP
    20  TOTL=TOTL+CB0(I)
        DO 21 I=1,NCOMP
    21  OI(I)=CB0(I)/TOTL
        CALL ISOIAS
        IF(JUDGE.NE.0)GO TO 43
        DG=(RHOP*Q0T*(1.0D+00-EBED))/(C0T*EBED)
        EON=6.0D+01/(TAU*DG)
        TSTEP=DSTEP*EON
        TOUT=TSTEP
        TTOTL=DTOTL*EON
        ST(3)=(1.0D+00-EBED)*TAU/(RP*EBED)
        EDS(3)=TAU*DG/(RP**2)
        EDP(3)=EDS(3)*C0T/(Q0T*RHOP)
        DO 22 I=1,NCOMP
         ST(I)=SKF(I)*ST(3)
         EDS(I)=DS(I)*EDS(3)
         EDP(I)=DP(I)*EDP(3)
         DOX(I)=3.0D+00*DG*ST(I)
    22 CONTINUE
        WRITE(7,680)US,TAU,EON,DG,C0T,Q0T
        WRITE(7,700) (CI(I),  I=1,NCOMP)
        WRITE(7,710) (ST(I),  I=1,NCOMP)
        WRITE(7,720) (EDS(I),I=1,NCOMP)
```

```
        IF(MODELD.EQ.2)WRITE(7,730)(EDP(I),I=1,NCOMP)
        ND=NDEG
        NND=NDEG*MC
        NE=NND+MC
        N1=NND+MC+NC
        N=N1*NCOMP
        NSTEP=0
        INDEX=1
        T0=0.0D+00
        DO 23 I=1,N
         Y0(I)=0.0D+00
  23    CONTINUE
        WRITE(7,690)
        IF(NCOMP.EQ.1)WRITE(7,740)
        IF(NCOMP.EQ.2)WRITE(7,750)
        IF(NCOMP.EQ.3)WRITE(7,760)
C
  30    CONTINUE
        ICOUNT=0
        CALL GEAR(N,T0,H0,Y0,TOUT,EPS,INDEX)
        IF(JUDGE.NE.0)GO TO 43
        TOTL=0.0D+00
        DO 31 I=1,NCOMP
         J=N1*I
         COUT(I)=Y0(J)
         IF(NRESLT.EQ.1)COUT(I)=COUT(I)*OI(I)/CI(I)
         TOTL=TOTL+COUT(I)
  31     CONTINUE
        T=TOUT/EON
        WRITE(1,770)ICOUNT,TOUT,T,(COUT(I),I=1,NCOMP),TOTL
        WRITE(7,770)ICOUNT,TOUT,T,(COUT(I),I=1,NCOMP),TOTL
        TOUT=TOUT+TSTEP
        NSTEP=NSTEP+1
        IF(TOUT.LT.TTOTL .AND. NSTEP.LE.200)GO TO 30
        TOI=0.0D+00
        TCI=0.0D+00
        DO 32 I=1,NCOMP
         TOI=TOI+OI(I)
         TCI=TCI+CI(I)
  32    CONTINUE
        IF(NRESLT.EQ.1)WRITE(7,780)(OI(I),I=1,NCOMP),TOI
        IF(NRESLT.EQ.2)WRITE(7,790)(CI(I),I=1,NCOMP),TCI
        GO TO 43
  40    CONTINUE
         WRITE(1,800)
         GO TO 43
  41    CONTINUE
         WRITE(1,810)
         GO TO 43
  42    CONTINUE
         WRITE(1,820)
  43    STOP
C
 100    FORMAT(/,8X,20A2,/,5(8X,I1,/))
 110    FORMAT(10(8X,D11.0,/))
 120    FORMAT(8X,3(D11.0,1X))
 130    FORMAT(8X,3(I1,11X))
 140    FORMAT(/,5X,I1)
 150    FORMAT(5X,I1)
 160    FORMAT(4F19.12)
 170    FORMAT(4F19.14)
```

```
  500 FORMAT(5X,'***** BREAKTHROUGH CURVE OF ',20A2,//,'========   INPUT
     + DATA   ========')
  510 FORMAT('MODELI :',I2,10X,'MODELD :',I2,10X,'NCOMP  :',I2,10X,'NDEG
     + :',I2)
  520 FORMAT('RP   :',D12.5,' RHOP :',D12.5,' DIA  :',D12.5,' BL   :'
     +,D12.5,//,'WT   :',D12.5,' FLRT :',D12.5,' DSTEP:',D12.5,' DTOTL
     +:',D12.5,//,'H0   :',D12.5,' EPS  :',D12.5)
  530 FORMAT('SKF(I)  :',3(D12.5,5X))
  540 FORMAT('CB0(I)  :',3(D12.5,5X))
  550 FORMAT('WM(I)   :',3(D12.5,5X))
  560 FORMAT('DS(I)   :',3(D12.5,5X))
  570 FORMAT('DP(I)   :',3(D12.5,5X))
  590 FORMAT('XK(I)   :',3(D12.5,5X))
  600 FORMAT('XN(I)   :',3(D12.5,5X))
  650 FORMAT('AU(I,J):')
  660 FORMAT('W(I):')
  670 FORMAT('B(I,J):')
  680 FORMAT('===== CALCULATION CONDITION =====',//,
     +'US   :',D12.5,' TAU  :',D12.5,' EON  :',D12.5,' DG   :',D12.5,
     +/,'COT  :',D12.5,' QOT  :',D12.5)
  690 FORMAT('===== BREAKTHROUGH CURVE  =====')
  700 FORMAT('CI(I)   :',3(D12.5,5X))
  710 FORMAT('ST(I)   :',3(D12.5,5X))
  720 FORMAT('EDS(I)  :',3(D12.5,5X))
  730 FORMAT('EDP(I)  :',3(D12.5,5X))
  740 FORMAT(6X,' TIME            TIME        C1/Co,t       TOTAL')
  750 FORMAT(6X,' TIME            TIME        C1/Co,t       C2/Co,t       TOTAL
     +')
  760 FORMAT(6X,' TIME            TIME        C1/Co,t       C2/Co,t       C3/Co,
     +t       TOTAL')
  770 FORMAT(I4,2X,F8.5,F10.3,4F13.7)
  780 FORMAT('AT EQUILIBRIUM(wt.fr.)  ',4F13.7)
  790 FORMAT('AT EQUILIBRIUM(mol.fr.)  ',4F13.7)
  800 FORMAT('USE COLM21.FTN FOR 3PM-MODEL')
  810 FORMAT('ERROR in "COLM2.DAT"')
  820 FORMAT('ISON(I) MUST BE 1')
      END
C
C
      SUBROUTINE ISOIAS
      IMPLICIT REAL*8(A-H,O-Z)
      INTEGER JUDGE,MODELD,NCOMP,IDUMMY
      INTEGER MX,I,L
      REAL*8 CB0,CI,WM,XK,XN,COT,QOT,DELTA
      REAL*8 SP,SH,E,ULIM,X1,X2,SPP,PAI,A
      COMMON /RTRN/ JUDGE,MODELD,NCOMP
      COMMON /SISO/ CB0(3),CI(3),WM(3)
      COMMON /SEQU/ XK(3),XN(3),COT,QOT,DELTA,IDUMMY(3)
      EXTERNAL FU1
      DATA DELTA,SP,SH,E,ULIM/1.0D-16,1.0D-04,2.0,1.0D-08,100.0/
      DATA MX/30/
C
      COT=0.0D+00
      DO 10 I=1,NCOMP
        X1=1.0D+00/(WM(I)*1.0D+03)
        CB0(I)=CB0(I)*X1
        COT=COT+CB0(I)
        X2=1.0D+00/XN(I)
        X2=(WM(I)*1.0D+03)**X2/WM(I)
        XK(I)=XK(I)*X2
   10 CONTINUE
```

```
      DO 11 I=1,NCOMP
   11 CI(I)=CB0(I)/COT
C
      SPP=SP
   12 CONTINUE
      CALL RFALSI(SPP,SH,E,PAI,FU1,ULIM,MX,L)
      IF(L.LT.MX .AND. PAI.LT.ULIM)GO TO 13
      IF(L.GE.MX)MX=MX+10
      IF(PAI.GE.ULIM)SPP=SPP/1.0D+02
      IF(MX.GE.100 .OR. SPP.LT.1.0D-18)GO TO 20
      GO TO 12
   13 CONTINUE
C
      PAI=PAI+DELTA
      QOT=0.0D+00
      DO 14 I=1,NCOMP
      A=XK(I)*XN(I)/PAI
      A=CB0(I)*(A**XN(I))
      QOT=QOT+A*XN(I)/PAI
   14 CONTINUE
      QOT=1.0D+00/QOT
      RETURN
C
   20 WRITE(1,500)MX,SPP
      JUDGE=1
      RETURN
  500 FORMAT('ERROR-500 in ISOIAS',/,'MX=',I3,'  SPP=',D15.7)
      END
C
C
      SUBROUTINE DIFFUN(Y0,F)
      IMPLICIT REAL*8(A-H,O-Z)
      INTEGER JUDGE,MODELD,NCOMP,MC,NC,ND,ICOUNT,NND,NE,N1
      INTEGER I,J,K,JC,II,JJ,KK,LL,NN
      REAL*8 Y0,F,CB0,CI,WM,AU,B,W,ST,EDS,EDP,DOX,DG,DUMMY
      REAL*8 AA,WW,BS,BP,YC,COUT
      DIMENSION Y0(126),F(126)
      DIMENSION AA(8),WW(8),BS(6,8),BP(6,8),YC(56,3),COUT(3)
      COMMON /RTRN/ JUDGE,MODELD,NCOMP
      COMMON /SISO/ CB0(3),CI(3),WM(3)
      COMMON /DIFF/ AU(8,8),B(7,7),W(7),ST(3),EDS(3),EDP(3),DOX(3),DG,MC
     +,NC,ND,ICOUNT
      COMMON /SEQU/ DUMMY(9),NND,NE,N1
C
      CALL EQUIAS(Y0,YC)
      DO 10 JC=1,NCOMP
      DO 11 J=1,MC
      WW(J)=0.0D+00
      AA(J)=0.0D+00
      DO 12 I=1,ND
      BS(I,J)=0.0D+00
      IF(MODELD.EQ.1)GO TO 12
      BP(I,J)=0.0D+00
   12 CONTINUE
   11 CONTINUE
C
      NN=N1*(JC-1)
      DO 20 I=1,ND
      DO 21 K=1,MC
      KK=NN+ND*(K-1)
      DO 22 J=1,ND
```

```
          JJ=KK+J
          BS(I,K)=BS(I,K)+B(I,J)*Y0(JJ)
          IF(MODELD.EQ.1)GO TO 22
          JJ=JJ-NN
          BP(I,K)=BP(I,K)+B(I,J)*YC(JJ,JC)
22    CONTINUE
21    CONTINUE
20 CONTINUE
      DO 23 K=1,MC
      J=NND+K
      II=NN+J
      JJ=NN+ND*(K-1)
      DO 24 I=1,ND
       KK=JJ+I
       F(KK)=EDS(JC)*(BS(I,K)+B(I,NC)*Y0(II))
       IF(MODELD.EQ.1)GO TO 25
       F(KK)=F(KK)+EDP(JC)*(BP(I,K)+B(I,NC)*YC(J,JC))
25    WW(K)=WW(K)+W(I)*F(KK)
24    CONTINUE
23 CONTINUE
C
      II=NND+1
      JJ=NN+II
      F(JJ)=(ST(JC)*(CI(JC)-YC(II,JC)) -WW(1))/W(NC)
      LL=NN+NND+NC
      DO 30 K=2,MC
       II=NND+K
       JJ=NN+II
       KK=NC+JJ
       F(JJ)=(ST(JC)*(Y0(KK)-YC(II,JC)) -WW(K))/W(NC)
       DO 31 J=2,MC
        KK=LL+J
        AA(K)=AA(K)+AU(K,J)*Y0(KK)
31    CONTINUE
30 CONTINUE
      LL=NN+NE
      DO 32 I=1,NC
       II=LL+I
       JJ=I+1
       KK=NND+JJ
       F(II)=DOX(JC)*(YC(KK,JC)-Y0(II)) -DG*(AA(JJ)+AU(JJ,1)*CI(JC))
32    CONTINUE
10 CONTINUE
      ICOUNT=ICOUNT+1
      KK=ICOUNT/200*200
      IF(KK.NE.ICOUNT)GO TO 40
       DO 41 I=1,NCOMP
       J=N1*I
       COUT(I)=Y0(J)
41    CONTINUE
      WRITE(1,500)ICOUNT,(COUT(I),I=1,NCOMP)
40 RETURN
500 FORMAT(I4,1X,3D15.7)
      END
C
C
      SUBROUTINE EQUIAS(Y0,YC)
      IMPLICIT REAL*8(A-H,O-Z)
      INTEGER JUDGE,MODELD,NCOMP,NND,NE,N1
      INTEGER LS,LC,KTOL,I,J
      REAL*8 Y0,YC,XK,XN,C0T,Q0T,DELTA
```

```
      REAL*8 QI,QT,SUM,A,DENO
      DIMENSION Y0(126),YC(56,3)
      DIMENSION QI(3)
      COMMON /RTRN/ JUDGE,MODELD,NCOMP
      COMMON /SEQU/ XK(3),XN(3),COT,QOT,DELTA,NND,NE,N1
C
      LS=1
      IF(MODELD.EQ.1)LS=NND+1
      DO 10 LC=LS,NE
       KTOL=0
       QT=0.0D+00
       DO 11 I=1,NCOMP
        J=N1*(I-1)+LC
        A=Y0(J)*QOT
        IF(A.GT.DELTA)GO TO 12
         QI(I)=0.0D+00
         YC(LC,I)=0.0D+00
         GO TO 11
   12   CONTINUE
        KTOL=1
        QI(I)=A
        QT=QT+A
   11  CONTINUE
       IF(KTOL.EQ.0)GO TO 10
C
       SUM=0.0D+00
       DO 13 I=1,NCOMP
        SUM=SUM+XN(I)*QI(I)
   13  CONTINUE
       DENO=QT*COT
       DO 14 I=1,NCOMP
        IF(QI(I).EQ.0.0D+00)GO TO 14
        A=SUM/(XK(I)*XN(I))
        YC(LC,I)=QI(I)*(A**XN(I))/DENO
   14  CONTINUE
   10 CONTINUE
      RETURN
      END
C
C
      SUBROUTINE RFALSI(SP,SH,E,X,FU,ULIM,MX,L)
      IMPLICIT REAL*8(A-H,O-Z)
      INTEGER MX,L
      REAL*8 SP,SH,E,X,ULIM
      REAL*8 B1,B2,D1,D2,F1,F2,H,TANG,X1,X2,XP
C
      X1=SP
      H=SH*DABS(X1)
      L=0
      D1=10.0*E
      D2=D1
      F1=FU(X1)
   11 X2=X1+H
      B1=X1
      B2=X2
      IF(X2.LT.ULIM)GO TO 12
      X=ULIM
      RETURN
   12 F2=FU(X2)
      IF(F1*F2.LE.0.0D+00)GO TO 13
      X1=X2
```

```
      F1=F2
      H=H*SH
      GO TO 11
   13 IF(DABS(D1).GE.E)GO TO 14
      X=X1
      RETURN
   14 IF(DABS(D2).GE.E)GO TO 15
      X=X2
      RETURN
   15 FP=F1
      XP=X1
      TANG=H/(F2-F1)
      X1=X1-TANG*F1
      F1=FU(X1)
      IF(F1*F2.LE.0.0D+00)GO TO 16
      F2=F1
      X2=X1
      F1=FP
      X1=XP
      D2=B2-X2
      IF(X2.NE.0.0D+00)D2=D2/X2
      B2=X2
      GO TO 17
   16 D1=B1-X1
      IF(X1.NE.0.0D+00)D1=D1/X1
      B1=X1
   17 H=X2-X1
      L=L+1
      IF(L.LT.MX)GO TO 13
      X=X1
      RETURN
      END
C
C
      REAL FUNCTION FU1*8(PAI)
      IMPLICIT REAL*8(A-H,O-Z)
      INTEGER IDUMMY,NCOMP,IDUM,I
      REAL*8 PAI,CB0,DUMMY,XK,XN,DUM,DELTA,SUM,X
      COMMON /RTRN/ IDUMMY(2),NCOMP
      COMMON /SISO/ CB0(3),DUMMY(6)
      COMMON /SEQU/ XK(3),XN(3),DUM(2),DELTA,IDUM(3)
C
      SUM=0.0D+00
      DO 10 I=1,NCOMP
       X=XK(I)*XN(I)/(PAI+DELTA)
       SUM=SUM+CB0(I)*X**XN(I)
   10 CONTINUE
      FU1=1.0D+00-SUM
      RETURN
      END
C                              September 20,1986  by C.Arai
C
C   =========    COMBINE "GEARCL.FTN"    =========
```

```
C          ************** FILE NAME : COLM21.FTN **************
C
C          This program gives the breakthrough curve for multi-
C          component systems. For the detail, a "Design of
C          Multi-Component Long Column Program COLM2.FTN Series"
C          is available upon request. September 20, 1986
C          by C. Arai in IIT, Chicago.
C
C          This program uses IAS and three-parameter model
C          for multi-component systems.
C
C            **********      FILE NAME : COLM21.FTN      **********
C
      IMPLICIT REAL*8(A-H,O-Z)
      INTEGER   JUDGE,MODELD,NCOMP,NND,NE,N1,ISON,IDUMMY,MODELI,MC,NC,ND,
     +ICOUNT
      INTEGER ITITLE,NDEG,NRESLT,NSTEP,INDEX,I,J,K,M,N,MM
      REAL*8 CB0,QB0,CI,COT,QOT,C0,XK,XN,SPR,WM,AI,BI,BETA,ETA,AU,B,W,ST
     +,EDS,EDP,DOX,DG,Q01
      REAL*8 Y0,SKF,DS,DP,COUT,RP,RHOP,DIA,BL,WT,FLRT,DSTEP,DTOTL,H0,EPS
     +,AREA,BEDVOL,EBED,US,TAU,EON,TSTEP,TOUT,TTOTL,T0,T,OI,TOTL,TOI,TCI
      DIMENSION Y0(126),SKF(3),DS(3),DP(3),COUT(3),OI(3),ITITLE(20)
      COMMON /RTRN/ JUDGE,MODELD,NCOMP
      COMMON /MAN1/ CB0(3),QB0(3)
      COMMON /MAN2/ CI(3),COT,QOT,NND,NE,N1
      COMMON /IAS1/ C0(3,3),XK(3,3),XN(3,3),Q01
      COMMON /IAS2/ WM(3),ISON(3),IDUMMY(3)
      COMMON /IAS3/ SPR(56,27)
      COMMON /M3PM/ AI(3),BI(3),BETA(3),ETA(3)
      COMMON /DIFF/ AU(8,8),B(7,7),W(7),ST(3),EDS(3),EDP(3),DOX(3),DG,MC
     +,MODELI,NC,ND,ICOUNT
C
      JUDGE=0
      READ(6,100)(ITITLE(I),I=1,20),MODELI,MODELD,NCOMP,NDEG,NRESLT
      READ(6,110)RP,RHOP,DIA,BL,WT,FLRT,DSTEP,DTOTL,H0,EPS
      READ(6,120)(SKF(I),I=1,NCOMP)
      READ(6,120)(CB0(I),I=1,NCOMP)
      READ(6,120)(WM(I),I=1,NCOMP)
      READ(6,120)(DS(I),I=1,NCOMP)
      READ(6,120)(DP(I),I=1,NCOMP)
      WRITE(7,500)(ITITLE(I),I=1,20)
      WRITE(7,510)MODELI,MODELD,NCOMP,NDEG
      WRITE(7,520)RP,RHOP,DIA,BL,WT,FLRT,DSTEP,DTOTL,H0,EPS
      WRITE(7,530)(SKF(I),I=1,NCOMP)
      WRITE(7,540)(CB0(I),I=1,NCOMP)
      WRITE(7,550)(WM(I),I=1,NCOMP)
      WRITE(7,560)(DS(I),I=1,NCOMP)
      WRITE(7,570)(DP(I),I=1,NCOMP)
      IF(MODELI.NE.1)GO TO 12
      READ(6,130)(ISON(I),I=1,NCOMP)
      M=1
      MM=1
      DO 10 I=1,NCOMP
       IF(ISON(I).GT.M)M=ISON(I)
       MM=MM*M
   10 CONTINUE
      DO 11 J=1,M
```

```
          IF(J.NE.1)READ(6,120) (CO(I,J),I=1,NCOMP)
          READ(6,120) (XK(I,J),I=1,NCOMP)
          READ(6,120) (XN(I,J),I=1,NCOMP)
          IF(J.NE.1)WRITE(7,580)J, (CO(I,J),I=1,NCOMP)
          WRITE(7,590)J, (XK(I,J),I=1,NCOMP)
          WRITE(7,600)J, (XN(I,J),I=1,NCOMP)
      11 CONTINUE
          GO TO 13
      12 CONTINUE
          IF(MODELI.NE.2)GO TO 40
          READ(6,120) (AI(I),   I=1,NCOMP)
          READ(6,120) (BI(I),   I=1,NCOMP)
          READ(6,120) (BETA(I),I=1,NCOMP)
          READ(6,120) (ETA(I),  I=1,NCOMP)
          WRITE(7,610) (AI(I)   ,I=1,NCOMP)
          WRITE(7,620) (BI(I)   ,I=1,NCOMP)
          WRITE(7,630) (BETA(I),I=1,NCOMP)
          WRITE(7,640) (ETA(I)  ,I=1,NCOMP)
C
      13 CONTINUE
          DO 14 K=2,NDEG
          IF(K.EQ.2)READ(5,140)N
          IF(K.NE.2)READ(5,150)N
          IF(K.NE.N)GO TO 41
          MC=K+2
          NC=K+1
          DO 15 I=1,MC
      15   READ(5,160) (AU(I,J),J=1,MC)
          READ(5,170) (W(I),I=1,NC)
          DO 16 I=1,NC
      16   READ(5,160) (B(I,J),J=1,NC)
      14 CONTINUE
C        WRITE(7,650)
C        DO 17 I=1,MC
C     17  WRITE(7,160) (AU(I,J),J=1,MC)
C        WRITE(7,660)
C        WRITE(7,170) (W(I),I=1,NC)
C        WRITE(7,670)
C        DO 18 I=1,NC
C     18  WRITE(7,160) (B(I,J),J=1,NC)
C
          AREA=3.14159D+00* (DIA**2)/4.0D+00
          BEDVOL=BL*AREA
          EBED= (BEDVOL-WT/RHOP)/BEDVOL
          US=FLRT/ (AREA*6.0D+01)
          TAU=EBED*BL/US
          TOTL=0.0D+00
          DO 20 I=1,NCOMP
      20 TOTL=TOTL+CB0(I)
          DO 21 I=1,NCOMP
      21 OI(I)=CB0(I)/TOTL
          IF(MODELI.EQ.1)CALL ISOIAS
          IF(MODELI.EQ.2)CALL ISO3PM
          IF(JUDGE.NE.0)GO TO 42
          DG=(RHOP*Q0T* (1.0D+00-EBED))/ (C0T*EBED)
          EON=6.0D+01/ (TAU*DG)
          TSTEP=DSTEP*EON
          TOUT=TSTEP
          TTOTL=DTOTL*EON
          ST(3)=(1.0D+00-EBED)*TAU/ (RP*EBED)
          EDS(3)=TAU*DG/ (RP**2)
          EDP(3)=EDS(3)*C0T/ (Q0T*RHOP)
          DO 22 I=1,NCOMP
```

```
        ST(I)=SKF(I)*ST(3)
        EDS(I)=DS(I)*EDS(3)
        EDP(I)=DP(I)*EDP(3)
        DOX(I)=3.0D+00*DG*ST(I)
   22 CONTINUE
        WRITE(7,680)US,TAU,EON,DG,COT,QOT
        WRITE(7,690)(QB0(I),I=1,NCOMP)
        WRITE(7,700)(CI(I),  I=1,NCOMP)
        WRITE(7,710)(ST(I),  I=1,NCOMP)
        WRITE(7,720)(EDS(I),I=1,NCOMP)
        IF(MODELD.EQ.2)WRITE(7,730)(EDP(I),I=1,NCOMP)
        ND=NDEG
        NND=NDEG*MC
        NE=NND+MC
        N1=NND+MC+NC
        N=N1*NCOMP
        NSTEP=0
        INDEX=1
        T0=0.0D+00
        DO 23 I=1,N
        Y0(I)=0.0D+00
   23 CONTINUE
        DO 24 I=1,NE
        DO 25 J=1,MM
   25   SPR(I,J)=1.0D-15
   24 CONTINUE
        IF(NCOMP.EQ.1)WRITE(7,740)
        IF(NCOMP.EQ.2)WRITE(7,750)
        IF(NCOMP.EQ.3)WRITE(7,760)
C
   30 CONTINUE
        ICOUNT=0
        CALL GEAR(N,T0,H0,Y0,TOUT,EPS,INDEX)
        IF(JUDGE.NE.0)GO TO 42
        TOTL=0.0D+00
        DO 31 I=1,NCOMP
        J=N1*I
        COUT(I)=Y0(J)
        IF(MODELI.EQ.1 .AND. NRESLT.EQ.1)COUT(I)=COUT(I)*OI(I)/CI(I)
        TOTL=TOTL+COUT(I)
   31   CONTINUE
        T=TOUT/EON
        WRITE(1,770)ICOUNT,TOUT,T,(COUT(I),I=1,NCOMP),TOTL
        WRITE(7,770)ICOUNT,TOUT,T,(COUT(I),I=1,NCOMP),TOTL
        TOUT=TOUT+TSTEP
        NSTEP=NSTEP+1
        IF(TOUT.LT.TTOTL .AND. NSTEP.LE.200)GO TO 30
        TOI=0.0D+00
        TCI=0.0D+00
        DO 32 I=1,NCOMP
        IF(MODELI.EQ.2)OI(I)=CI(I)
        TOI=TOI+OI(I)
        TCI=TCI+CI(I)
   32 CONTINUE
        IF(NRESLT.EQ.1)WRITE(7,780)(OI(I),I=1,NCOMP),TOI
        IF(NRESLT.EQ.2)WRITE(7,790)(CI(I),I=1,NCOMP),TCI
        GO TO 42
   40 CONTINUE
        WRITE(1,800)
        GO TO 42
   41 CONTINUE
        WRITE(1,810)
   42 STOP
```

```
C
  100 FORMAT(/,8X,20A2,/,5(8X,I1,/))
  110 FORMAT(10(8X,D11.0,/))
  120 FORMAT(8X,3(D11.0,1X))
  130 FORMAT(8X,3(I1,11X))
  140 FORMAT(/,5X,I1)
  150 FORMAT(5X,I1)
  160 FORMAT(4F19.12)
  170 FORMAT(4F19.14)
  500 FORMAT(5X,'***** BREAKTHROUGH CURVE OF ',20A2,//,'======  INPUT
     +  DATA   =======')
  510 FORMAT('MODELI :',I2,10X,'MODELD :',I2,10X,'NCOMP  :',I2,10X,'NDEG
     +  :',I2)
  520 FORMAT('RP    :',D12.5,' RHOP :',D12.5,' DIA :',D12.5,' BL   :'
     +,D12.5,/,'WT    :',D12.5,'  FLRT :',D12.5,'  DSTEP:',D12.5,' DTOTL
     +:',D12.5,/,'H0    :',D12.5,' EPS  :',D12.5)
  530 FORMAT('SKF(I)   :',3(D12.5,5X))
  540 FORMAT('CB0(I)   :',3(D12.5,5X))
  550 FORMAT('WM(I)    :',3(D12.5,5X))
  560 FORMAT('DS(I)    :',3(D12.5,5X))
  570 FORMAT('DP(I)    :',3(D12.5,5X))
  580 FORMAT('C0(I,',I1,') :',3(D12.5,5X))
  590 FORMAT('XK(I,',I1,') :',3(D12.5,5X))
  600 FORMAT('XN(I,',I1,') :',3(D12.5,5X))
  610 FORMAT('AI(I)    :',3(D12.5,5X))
  620 FORMAT('BI(I)    :',3(D12.5,5X))
  630 FORMAT('BETA(I) :',3(D12.5,5X))
  640 FORMAT('ETA(I)  :',3(D12.5,5X))
  650 FORMAT('AU(I,J):')
  660 FORMAT('W(I):')
  670 FORMAT('B(I,J):')
  680 FORMAT('===== CALCULATION CONDITION =====',//,
     +'US   :',D12.5,' TAU :',D12.5,' EON :',D12.5,' DG   :',D12.5,
     +/,'COT  :',D12.5,' QOT :',D12.5)
  690 FORMAT('QB0(I)   :',3(D12.5,5X))
  700 FORMAT('CI(I)    :',3(D12.5,5X))
  710 FORMAT('ST(I)    :',3(D12.5,5X))
  720 FORMAT('EDS(I)   :',3(D12.5,5X))
  730 FORMAT('EDP(I)   :',3(D12.5,5X))
  740 FORMAT(6X,'  TIME         TIME       C1/Co,t       TOTAL')
  750 FORMAT(6X,'  TIME         TIME       C1/Co,t       C2/Co,t       TOTAL
     +')
  760 FORMAT(6X,'  TIME         TIME       C1/Co,t       C2/Co,t       C3/Co,
     +t       TOTAL')
  770 FORMAT(I4,2X,F8.5,F10.3,4F13.7)
  780 FORMAT('AT EQUILIBRIUM(wt.fr.)  ',4F13.7)
  790 FORMAT('AT EQUILIBRIUM(mol.fr.) ',4F13.7)
  800 FORMAT('ERROR in "COLM2-IAS" or "COLM2-3PM"')
  810 FORMAT('ERROR in "COLM2.DAT"')
      END
C
C

      SUBROUTINE ISOIAS
      IMPLICIT REAL*8(A-H,O-Z)
      INTEGER JUDGE,MODELD,NCOMP,IDUMMY,ISON,K1E,K2E,K3E,KSEC,IDUM
      INTEGER KDIM,MX,I,J,K,L,JJ,KK,K1,K2,K3
      REAL*8 CB0,QB0,CI,C0T,Q0T,C0,XK,XN,Q01,WM,DELTA,QI
      REAL*8 SDIM,ZB0,SP,SH,E,ULIM,BLIM,X1,X2,SUM,PAI,CI0,SPP
      DIMENSION SDIM(3,3),ZB0(3),KDIM(3)
      COMMON /RTRN/ JUDGE,MODELD,NCOMP
```

```
      COMMON /MAN1/ CB0(3),QB0(3)
      COMMON /MAN2/ CI(3),COT,QOT,IDUMMY(3)
      COMMON /IAS1/ CO(3,3),XK(3,3),XN(3,3),Q01
      COMMON /IAS2/ WM(3),ISON(3),K1E,K2E,K3E
      COMMON /FUN1/ DELTA(3,3),QI(3),KSEC(3),IDUM(4)
      EXTERNAL FU1
      DATA Q01,SP,SH,E,ULIM,BLIM/1.0D-16,1.0D-04,2.0,1.0D-08,100.0,0.0/
      DATA MX/50/
C
      COT=0.0D+00
      DO 10 I=1,NCOMP
       X1=1.0D+00/(WM(I)*1.0D+03)
       CB0(I)=CB0(I)*X1
       QI(I)=CB0(I)
       COT=COT+CB0(I)
       J=ISON(I)
       DO 11 K=1,J
        IF(J.NE.1)CO(I,K)=CO(I,K)*X1
        X2=1.0D+00/XN(I,K)
        X2=(WM(I)*1.0D+03)**X2/WM(I)
        XK(I,K)=XK(I,K)*X2
   11  CONTINUE
       X1=XK(I,1)*(XN(I,1)-1.0D+00)
       IF(X1.LE.0.0D+00)GO TO 53
       CO(I,1)=(Q01*2.0D+00/X1)**XN(I,1)
   10 CONTINUE
      DO 12 I=1,NCOMP
   12 CI(I)=CB0(I)/COT
C
      DO 20 I=1,NCOMP
       L=ISON(I)
       DO 21 K=1,L
        X1=1.0D+00/XN(I,K)
        X1=CO(I,K)**X1
        SUM=XK(I,K)*XN(I,K)*X1
        IF(K.EQ.1)GO TO 23
        KK=K-1
        DO 22 J=1,KK
         X1=1.0D+00/XN(I,J)
         JJ=J+1
         X1=CO(I,JJ)**X1-CO(I,J)**X1
         SUM=SUM-XK(I,J)*XN(I,J)*X1
   22   CONTINUE
   23   CONTINUE
        X1=1.0D+00/XN(I,1)
        X1=XK(I,1)*(1.0D+00+XN(I,1))*CO(I,1)**X1/2.0D+00
        DELTA(I,K)=SUM-X1
        IF(DELTA(I,K).LE.0.0D+00)GO TO 52
   21  CONTINUE
C      WRITE(7,540)(DELTA(I,K),K=1,L)
   20 CONTINUE
C
      IF(NCOMP.EQ.3)GO TO 31
       J=NCOMP+1
       DO 30 I=J,3
   30   ISON(I)=1
   31 CONTINUE
      K3E=ISON(3)
      K2E=ISON(2)
      K1E=ISON(1)
      DO 32 K3=1,K3E
       KSEC(3)=K3
```

```
      DO 33 K2=1,K2E
      KSEC(2)=K2
      DO 34 K1=1,K1E
      KSEC(1)=K1
      SPP=SP
38    CONTINUE
      CALL RFALSI(SPP,SH,E,PAI,FU1,ULIM,BLIM,MX,L)
      IF(L.LT.MX .AND. PAI.LT.ULIM)GO TO 37
      IF(L.GE.MX)MX=MX+10
      SPP=SPP/1.0D+02
      IF(MX.GE.100 .OR. SPP.LT.1.0D-16)GO TO 50
      GO TO 38
37    CONTINUE
      DO 35 I=1,NCOMP
      K=KSEC(I)
      CI0=(PAI+DELTA(I,K))/(XK(I,K)*XN(I,K))
      SDIM(I,K)=1.0D+00/CI0
      CI0=CI0**XN(I,K)
      IF(K.EQ.ISON(I))GO TO 36
      IF(CI0.LT.C0(I,K))GO TO 34
      J=K+1
      IF(CI0.GT.C0(I,J))GO TO 34
36    KDIM(I)=K
35    CONTINUE
      GO TO 39
34    CONTINUE
33   CONTINUE
32  CONTINUE
     GO TO 51
39 CONTINUE
C
      Q0T=0.0D+00
      DO 40 I=1,NCOMP
      K=KDIM(I)
      ZB0(I)=CB0(I)*(SDIM(I,K)**XN(I,K))
      Q0T=Q0T+(ZB0(I)*XN(I,K))/(PAI+DELTA(I,K))
40    CONTINUE
      Q0T=1.0D+00/Q0T
      DO 41 I=1,NCOMP
      QB0(I)=Q0T*ZB0(I)
41    CONTINUE
      RETURN
C
50   WRITE(1,500)
      JUDGE=1
      RETURN
51   WRITE(1,510)
      JUDGE=1
      RETURN
52   WRITE(1,520)
      JUDGE=1
      RETURN
53   WRITE(1,530)
      JUDGE=1
      RETURN
500 FORMAT('ERROR-500 in ISOIAS')
510 FORMAT('ERROR-510 in ISOIAS')
520 FORMAT('ERROR-520 in ISOIAS')
530 FORMAT('ERROR-530 in ISOIAS')
540 FORMAT('DELTA(I):',4(D12.5,5X))
      END
C
C
```

```
      SUBROUTINE ISO3PM
      IMPLICIT REAL*8(A-H,O-Z)
      INTEGER IDUM,NCOMP,IDUMMY,I
      REAL*8 CB0,QB0,CI,COT,QOT,AI,BI,BETA,ETA,DENO
      COMMON /RTRN/ IDUM(2),NCOMP
      COMMON /MAN1/ CB0(3),QB0(3)
      COMMON /MAN2/ CI(3),COT,QOT,IDUMMY(3)
      COMMON /M3PM/ AI(3),BI(3),BETA(3),ETA(3)
C
      COT=0.0D+00
      DENO=1.0D-03
      DO 10 I=1,NCOMP
       CB0(I)=CB0(I)*DENO
       COT=COT+CB0(I)
       AI(I)=AI(I)/(ETA(I)*DENO)
       BI(I)=BI(I)/(ETA(I)*DENO)**BETA(I)
   10 CONTINUE
      DO 11 I=1,NCOMP
   11  CI(I)=CB0(I)/COT
      DENO=1.0D+00
      DO 12 I=1,NCOMP
   12 DENO=DENO+BI(I)*CB0(I)**BETA(I)
      QOT=0.0D+00
      DO 13 I=1,NCOMP
       QB0(I)=AI(I)*CB0(I)/DENO
       QOT=QOT+QB0(I)
   13 CONTINUE
      RETURN
      END
C
C
      SUBROUTINE DIFFUN(Y0,F)
      IMPLICIT REAL*8(A-H,O-Z)
      INTEGER JUDGE,MODELD,NCOMP,NND,NE,N1,MC,MODELI,NC,ND,ICOUNT
      INTEGER I,J,K,JC,II,JJ,KK,LL,NN
      REAL*8 Y0,F,CI,DUMMY,AU,B,W,ST,EDS,EDP,DOX,DG
      REAL*8 AA,WW,BS,BP,YC,COUT
      DIMENSION Y0(126),F(126)
      DIMENSION AA(8),WW(8),BS(6,8),BP(6,8),YC(56,3),COUT(3)
      COMMON /RTRN/ JUDGE,MODELD,NCOMP
      COMMON /MAN2/ CI(3),DUMMY(2),NND,NE,N1
      COMMON /DIFF/ AU(8,8),B(7,7),W(7),ST(3),EDS(3),EDP(3),DOX(3),DG,MC
     +,MODELI,NC,ND,ICOUNT
C
      IF(MODELI.EQ.1)CALL EQUIAS(Y0,YC)
      IF(MODELI.EQ.2)CALL EQU3PM(Y0,YC)
      IF(JUDGE.NE.0)GO TO 40
      DO 10 JC=1,NCOMP
      DO 11 J=1,MC
       WW(J)=0.0D+00
       AA(J)=0.0D+00
       DO 12 I=1,ND
        BS(I,J)=0.0D+00
        IF(MODELD.EQ.1)GO TO 12
        BP(I,J)=0.0D+00
   12   CONTINUE
   11 CONTINUE
C
      NN=N1*(JC-1)
      DO 20 I=1,ND
       DO 21 K=1,MC
        KK=NN+ND*(K-1)
```

```
          DO 22 J=1,ND
          JJ=KK+J
          BS(I,K)=BS(I,K)+B(I,J)*Y0(JJ)
          IF(MODELD.EQ.1)GO TO 22
          JJ=JJ-NN
          BP(I,K)=BP(I,K)+B(I,J)*YC(JJ,JC)
  22      CONTINUE
  21    CONTINUE
  20  CONTINUE
        DO 23 K=1,MC
        J=NND+K
        II=NN+J
        JJ=NN+ND*(K-1)
        DO 24 I=1,ND
        KK=JJ+I
        F(KK)=EDS(JC)*(BS(I,K)+B(I,NC)*Y0(II))
        IF(MODELD.EQ.1)GO TO 25
        F(KK)=F(KK)+EDP(JC)*(BP(I,K)+B(I,NC)*YC(J,JC))
  25    WW(K)=WW(K)+W(I)*F(KK)
  24    CONTINUE
  23  CONTINUE
C
        II=NND+1
        JJ=NN+II
        F(JJ)=(ST(JC)*(CI(JC)-YC(II,JC))  -WW(1))/W(NC)
        LL=NN+NND+NC
        DO 30 K=2,MC
        II=NND+K
        JJ=NN+II
        KK=NC+JJ
        F(JJ)=(ST(JC)*(Y0(KK)-YC(II,JC))  -WW(K))/W(NC)
        DO 31 J=2,MC
        KK=LL+J
        AA(K)=AA(K)+AU(K,J)*Y0(KK)
  31    CONTINUE
  30  CONTINUE
        LL=NN+NE
        DO 32 I=1,NC
        II=LL+I
        JJ=I+1
        KK=NND+JJ
        F(II)=DOX(JC)*(YC(KK,JC)-Y0(II))  -DG*(AA(JJ)+AU(JJ,1)*CI(JC))
  32  CONTINUE
  10  CONTINUE
        ICOUNT=ICOUNT+1
        KK=ICOUNT/200*200
        IF(KK.NE.ICOUNT)GO TO 40
        DO 41 I=1,NCOMP
        J=N1*I
        COUT(I)=Y0(J)
  41    CONTINUE
        WRITE(1,500)ICOUNT,(COUT(I),I=1,NCOMP)
  40  RETURN
 500  FORMAT(I4,1X,3D15.7)
        END
C
C
        SUBROUTINE EQUIAS(Y0,YC)
        IMPLICIT REAL*8(A-H,O-Z)
        INTEGER JUDGE,MODELD,NCOMP,NND,NE,N1,ISON,K1E,K2E,K3E,KSEC,NEG,
       +KTOL
        INTEGER MX,I,J,K,L,LC,LS,K1,K2,K3,KK,LL
```

```
      REAL*8 YO,YC,CI,COT,QOT,CO,XK,XN,Q01,SPR,WM,DELTA,QI
      REAL*8 SH,E,ULIM,BLIM,AA,BB,CC,SA,SB,SC,SD,PAI,QT,SP
      DIMENSION YO(126),YC(56,3)
      COMMON /RTRN/ JUDGE,MODELD,NCOMP
      COMMON /MAN2/ CI(3),COT,QOT,NND,NE,N1
      COMMON /IAS1/ CO(3,3),XK(3,3),XN(3,3),Q01
      COMMON /IAS2/ WM(3),ISON(3),K1E,K2E,K3E
      COMMON /IAS3/ SPR(56,27)
      COMMON /FUN1/ DELTA(3,3),QI(3),KSEC(3),NEG(3),KTOL
      EXTERNAL FU2
      DATA SH,E,ULIM,BLIM/2.0,1.0D-07,100.0,0.0/
      DATA MX/50/
C
      LS=1
      IF(MODELD.EQ.1)LS=NND+1
      DO 10 LC=LS,NE
       QT=0.0D+00
       KTOL=0
      DO 11 I=1,NCOMP
       J=N1*(I-1)+LC
       AA=YO(J)*QOT
       IF(AA.GT.Q01)GO TO 12
        YC(LC,I)=0.0D+00
        GO TO 11
   12 CONTINUE
       QI(I)=AA
       QT=QT+AA
       KTOL=KTOL+1
       NEG(KTOL)=I
   11 CONTINUE
      IF(KTOL.EQ.0)GO TO 10
C
      KK=0
      DO 20 K3=1,K3E
       KSEC(3)=K3
       DO 21 K2=1,K2E
        KSEC(2)=K2
        DO 22 K1=1,K1E
         KSEC(1)=K1
         GO TO (23,24,25),KTOL
   23    CONTINUE
          LL=NEG(1)
          K=KSEC(LL)
          PAI=XN(LL,K)*QI(LL) -DELTA(LL,K)
          IF(PAI.LT.0.0D+00)GO TO 33
          GO TO 27
   24    CONTINUE
          LL=NEG(1)
          K=KSEC(LL)
          SA=DELTA(LL,K)
          SB=XN(LL,K)*QI(LL)
          LL=NEG(2)
          K=KSEC(LL)
          SC=DELTA(LL,K)
          SD=XN(LL,K)*QI(LL)
          BB=SA+SC-(SB+SD)
          CC=SA*SC -SB*SC -SA*SD
          CC=BB**2 -4.0D+00*CC
          IF(CC.LT.0.0D+00)GO TO 30
          PAI=(DSQRT(CC)-BB)/(2.0D+00)
          IF(PAI.GT.Q01)GO TO 27
          SP=1.0D-17
          KK=1
```

```
              GO TO 26
      25  CONTINUE
              KK=KK+1
              SP=SPR(LC,KK)
      26  CONTINUE
              CALL RFALSI(SP,SH,E,PAI,FU2,ULIM,BLIM,MX,L)
              SPR(LC,KK)=PAI/1.0D+01
              IF(PAI.LT.ULIM .AND. L.LT.MX)GO TO 27
              IF(L.GE.MX)MX=MX+10
              IF(PAI.GE.ULIM)SP=SP/1.0D+02
              IF(MX.GE.100 .OR. SP.LT.1.0D-20)GO TO 31
              GO TO 26
      27  CONTINUE
              DO 29 LL=1,KTOL
              I=NEG(LL)
              K=KSEC(I)
              AA=(PAI+DELTA(I,K))/(XK(I,K)*XN(I,K))
              AA=QI(I)*AA**XN(I,K)/QT
              IF(K.EQ.ISON(I))GO TO 28
              IF(K.NE.1 .AND. AA.LT.CO(I,K))GO TO 22
              J=K+1
              IF(AA.GT.CO(I,J))GO TO 22
      28  YC(LC,I)=AA/COT
      29  CONTINUE
              GO TO 10
      22  CONTINUE
      21   CONTINUE
      20  CONTINUE
              GO TO 32
      10  CONTINUE
              RETURN
C
      30  WRITE(1,500)
              JUDGE=1
              RETURN
      31  WRITE(1,510)L,MX,SPR
              JUDGE=1
              RETURN
      32  WRITE(1,520)
              JUDGE=1
              RETURN
      33  WRITE(1,530)PAI
              JUDGE=1
              RETURN
      500 FORMAT('ERROR-500 IN EQUIAS')
      510 FORMAT('ERROR-510 IN EQUIAS',/,'L=',I3,' MX=',I3,' SPR=',D15.7)
      520 FORMAT('ERROR-520 IN EQUIAS')
      530 FORMAT('ERROR-530 IN EQUIAS',/,'PAI=',D25.16)
              END
C
C
              SUBROUTINE EQU3PM(Y0,YC)
              IMPLICIT REAL*8(A-H,O-Z)
              INTEGER JUDGE,MODELD,NCOMP,NND,NE,N1,NEG,KTOL
              INTEGER MX,LS,LC,I,J,I1,II,L
              REAL*8 Y0,YC,CI,COT,QOT,AI,BI,BETA,ETA,QI,ALPHA
              REAL*8 CUTL,RLIM,SH,E,ULIM,BLIM,SP,BL,CP
              DIMENSION Y0(126),YC(56,3)
              COMMON /RTRN/ JUDGE,MODELD,NCOMP
              COMMON /MAN2/ CI(3),COT,QOT,NND,NE,N1
              COMMON /M3PM/ AI(3),BI(3),BETA(3),ETA(3)
              COMMON /FUN2/ QI(3),ALPHA(3),NEG(3),KTOL
              EXTERNAL FU3,FU4
```

```
      DATA CUTL,RLIM/1.0D-15,1.0D-24/
      DATA SH,E,ULIM,BLIM/0.5,1.0D-07,100.0,1.0D-16/
      DATA MX/50/
C
      LS=1
      IF(MODELD.EQ.1)LS=NND+1
      DO 100 LC=LS,NE
       KTOL=0
       DO 10 I=1,NCOMP
        J=N1*(I-1)+LC
        IF(Y0(J).LT.CUTL)GO TO 11
         KTOL=KTOL+1
         NEG(KTOL)=I
         QI(KTOL)=Y0(J)*Q0T
         GO TO 10
   11    YC(LC,I)=0.0D+00
   10  CONTINUE
       IF(KTOL.EQ.0)GO TO 100
C
       SP=QI(1)/1.0D+02
       BL=BLIM
       IF(KTOL.NE.1)GO TO 30
   20  CONTINUE
       CALL RFALSI(SP,SH,E,CP,FU3,ULIM,BL,MX,L)
       IF(L.GE.MX .OR. CP.LE.BL)GO TO 21
        I=NEG(1)
        YC(LC,I)=CP/C0T
        GO TO 100
   21  IF(L.LT.MX)GO TO 22
        MX=MX+10
        WRITE(1,520)LC,KTOL,L,BL,SP,CP
        GO TO 23
   22  BL=BL/1.0D+01
       SP=SP*1.0D+01
       WRITE(1,530)LC,KTOL,L,BL,SP,CP
   23  IF(BL.LT.RLIM .OR. MX.GE.100)GO TO 40
       GO TO 20
C
   30  CONTINUE
       I1=NEG(1)
       DO 31 I=2,KTOL
        II=NEG(I)
        ALPHA(II)=QI(II)*AI(I1)/(QI(I1)*AI(II))
   31  CONTINUE
   32  CONTINUE
       CALL RFALSI(SP,SH,E,CP,FU4,ULIM,BL,MX,L)
       IF(L.GE.MX .OR. CP.LT.BL)GO TO 34
        YC(LC,I1)=CP/C0T
        DO 33 I=2,KTOL
         II=NEG(I)
         YC(LC,II)=CP*ALPHA(II)/C0T
   33   CONTINUE
        GO TO 100
   34  IF(L.LT.MX)GO TO 35
        MX=MX+10
        WRITE(1,520)LC,KTOL,L,BL,SP,CP
        GO TO 36
   35  BL=BL/1.0D+01
       SP=SP*1.0D+01
       WRITE(1,530)LC,KTOL,L,BL,SP,CP
   36  IF(BL.LT.RLIM .OR. MX.GE.100)GO TO 41
```

```
          GO TO 32
    100 CONTINUE
          RETURN
C
     40 WRITE(1,500)L,BL,CP
          JUDGE=1
          RETURN
     41 WRITE(1,510)L,BL,CP
          JUDGE=1
          RETURN
    500 FORMAT('ERROR-500 IN EQU3PM',/,
         +'L=',I3,' BL=',D15.7,' CP=',D15.7)
    510 FORMAT('ERROR-510 IN EQU3PM',/,
         +'L=',I3,' BL=',D15.7,' CP=',D15.7)
    520 FORMAT('MX increased. This is no warning!',/,3I4,3D15.7)
    530 FORMAT('SP increased. This is no warning!',/,3I4,3D15.7)
          END
C
C
          SUBROUTINE RFALSI(SP,SH,E,X,FU,ULIM,BLIM,MX,L)
          IMPLICIT REAL*8(A-H,O-Z)
          INTEGER MX,L
          REAL*8 SP,SH,E,X,ULIM,BLIM
          REAL*8 B1,B2,D1,D2,F1,F2,H,TANG,X1,X2,XP
C
          X1=SP
          H=SH*DABS(X1)
          L=0
          D1=10.0*E
          D2=D1
          F1=FU(X1)
     10 X2=X1+H
          IF(SH.LT.1.0D+00)X2=H
          B1=X1
          B2=X2
          IF(X2.GT.BLIM)GO TO 11
          X=BLIM
          RETURN
     11 IF(X2.LT.ULIM)GO TO 12
          X=ULIM
          RETURN
     12 F2=FU(X2)
          IF(F1*F2.LE.0.0D+00)GO TO 13
          X1=X2
          F1=F2
          H=H*SH
          GO TO 10
     13 IF(DABS(D1).GE.E)GO TO 14
          X=X1
          RETURN
     14 IF(DABS(D2).GE.E)GO TO 15
          X=X2
          RETURN
     15 FP=F1
          XP=X1
          TANG=H/(F2-F1)
          X1=X1-TANG*F1
          F1=FU(X1)
          IF(F1*F2.LE.0.0D+00)GO TO 16
          F2=F1
          X2=X1
```

```
      F1=FP
      X1=XP
      D2=B2-X2
      IF(X2.NE.0.0D+00)D2=D2/X2
      B2=X2
      GO TO 17
   16 D1=B1-X1
      IF(X1.NE.0.0D+00)D1=D1/X1
      B1=X1
   17 H=X2-X1
      L=L+1
      IF(L.LT.MX)GO TO 13
      X=X1
      IF(DABS(F2).LT.DABS(F1))X=X2
      RETURN
      END
C
C

      REAL FUNCTION FU1*8(PAI)
      IMPLICIT REAL*8(A-H,O-Z)
      INTEGER IDUMMY,NCOMP,KSEC,IDUM,I,K
      REAL*8 PAI,C0,XK,XN,Q01,DELTA,QI,SUM,X
      COMMON /RTRN/ IDUMMY(2),NCOMP
      COMMON /IAS1/ C0(3,3),XK(3,3),XN(3,3),Q01
      COMMON /FUN1/ DELTA(3,3),QI(3),KSEC(3),IDUM(4)
C
      SUM=0.0D+00
      DO 10 I=1,NCOMP
       K=KSEC(I)
       X=XK(I,K)*XN(I,K)/(PAI+DELTA(I,K))
       SUM=SUM+QI(I)*X**XN(I,K)
   10 CONTINUE
      FU1=1.0D+00-SUM
      RETURN
      END
C
C

      REAL FUNCTION FU2*8(PAI)
      IMPLICIT REAL*8(A-H,O-Z)
      INTEGER KSEC,NEG,KTOL,I,J,K
      REAL*8 PAI,C0,XK,XN,Q01,DELTA,QI,SUM
      COMMON /IAS1/ C0(3,3),XK(3,3),XN(3,3),Q01
      COMMON /FUN1/ DELTA(3,3),QI(3),KSEC(3),NEG(3),KTOL
C
      SUM=0.0D+00
      DO 10 J=1,KTOL
       I=NEG(J)
       K=KSEC(I)
       SUM=SUM+(XN(I,K)*QI(I))/(PAI+DELTA(I,K))
   10 CONTINUE
      FU2=SUM-1.0D+00
      RETURN
      END
C
C

      REAL FUNCTION FU3*8(CP)
      IMPLICIT REAL*8(A-H,O-Z)
      INTEGER NEG,KTOL,I
      REAL*8 CP,AI,BI,BETA,ETA,QI,ALPHA,DENO
      COMMON /M3PM/ AI(3),BI(3),BETA(3),ETA(3)
      COMMON /FUN2/ QI(3),ALPHA(3),NEG(3),KTOL
```

```
C
      I=NEG(1)
      DENO=1.0D+00+BI(I)*CP**BETA(I)
      FU3=QI(I)*DENO-AI(I)*CP
      RETURN
      END
C
C
      REAL FUNCTION FU4*8(CP)
      IMPLICIT REAL*8(A-H,O-Z)
      INTEGER NEG,KTOL,I,J,K
      REAL*8 CP,AI,BI,BETA,ETA,QI,ALPHA,X
      COMMON /M3PM/ AI(3),BI(3),BETA(3),ETA(3)
      COMMON /FUN2/ QI(3),ALPHA(3),NEG(3),KTOL
C
      I=NEG(1)
      X=1.0D+00+BI(I)*CP**BETA(I)
      DO 10 J=2,KTOL
       K=NEG(J)
       X=X+BI(K)*(ALPHA(K)*CP)**BETA(K)
   10 CONTINUE
      FU4=QI(I)*X-AI(I)*CP
      RETURN
      END
C                           September 20,1986 by C.Arai
C
C     ==========     COMBINE "GEARCL.FTN"     ==========
```

# INDEX